Heinz Bloch

Fluid Machinery

Also of interest

Plant Engineering.
A Guide for the Construction of Process Plants
Lehner, Kepplinger, Friedacher, 2021
ISBN 978-3-11-057330-5, e-ISBN 978-3-11-057332-9

Process Engineering.
Addressing the Gap between Study and Chemical Industry
Kleiber, 2020
ISBN 978-3-11-065764-7, e-ISBN 978-3-11-065768-5

Microfluidics.
Applications in Biotechnology
Andar, Adelstein, Adelstein, 2020
ISBN 978-3-11-063615-4, e-ISBN 978-3-11-063633-8

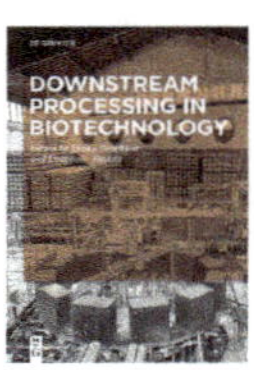

Downstream Processing in Biotechnology
Beschkov, Yankov (Eds.), 2020
ISBN 978-3-11-057395-4, e-ISBN 978-3-11-057411-1

Falling Films in Desalination.
A Computational Approach
Raach, 2019
ISBN 978-3-11-059177-4, e-ISBN 978-3-11-059233-7

Heinz Bloch

Fluid Machinery

Life Extension of Pumps, Gas Compressors and Drivers

DE GRUYTER

Author
Heinz Bloch, P. E.
267 Sunnyvale East
Montgomery, TX 77356
USA
heinzpbloch@gmail.com

ISBN 978-3-11-067413-2
e-ISBN (PDF) 978-3-11-067415-6
e-ISBN (EPUB) 978-3-11-067427-9

Library of Congress Control Number: 2020933147

Bibliographic information published by the Deutsche Nationalbibliothek
The Deutsche Nationalbibliothek lists this publication in the Deutsche Nationalbibliografie; detailed bibliographic data are available on the Internet at http://dnb.dnb.de.

Cover image: MAN, Diesel-Borsig, Berlin, Germany
Typesetting: Integra Software Services Pvt. Ltd.
Printing and binding: CPI books GmbH, Leck

www.degruyter.com

Foreword

In about 2000, an experienced equipment reliability engineer and Technical Manager[1] explained how he had made good use of the author's first book, a text published in 1982 with the title *Improving Machinery Reliability*. Apparently, he found it helpful to examine, and then implemented, some of the book's stated principles and recommendations. He had generously called them guiding elements that allowed him to write better specifications, take a more active role in vendor selection, and getting involved in the manufacturers' process of building and installing these machines at his plant. When we met again a few years later, we re-visited his successes and asked each other how these could be incorporated in a fourth edition of the 1982 book.

However, the third edition released in 1998 [1] had filled 680 pages and, as 2020 approached, the reading habits of young engineers had moved in the direction of acquiring slimmer books. The author's focus – since 2011 – had also been on sharing important information in readable hard-cover books of perhaps 210–230 pages. Although generally agreeing with this page count, many publishers had reached the conclusion that paperback books with ever less expensive covers were the way to go. But, in the author's opinion, such covers would not be consistent with the quality and permanence sought by a reading audience made up of productive reliability professionals.

It was time to remind ourselves (and like-minded readers) that "only dead fish *always* swim downstream with the flow." Yes, some of us still cling to the belief that doing the right thing is more like a journey that occasionally requires us to swim against the common tide. In this instance, higher-quality cover options had to be identified and introduced to a receptive publishing company.

Fortunately, much can be said in about 230 pages, and even the new title *Fluid Machines: Improving the Life of Pumps, Compressors and Turbines* denotes a bit of a break with the past. As the author and his editor (using "we" because a competent editor was involved) set out to prove that value is sometimes found in fewer words, both of us agreed on this shorter, condensed, crisp, totally revised, and fully updated text. It finally materialized, many years after the experienced equipment reliability engineer-manager had given his favorable comments. The present text will again give guidance on how to cost-justify and upgrade aging machines, and how to specify new machines with greater forethought.

That said, however, it is not within the scope of this book to describe fluid machines in much detail. Many books, some by the author of [2] and [3], have described basic designs quite well since about 1980. Machinery reliability almost always involves lubrication, and we point to [4] without hesitation. But, instead of repeating the contents of these four references, we want to acquaint reliability professionals,

1 Robert DeMaria was a respected engineer and colleague. He succumbed to a vicious form of cancer in early 2015, aged 67. We honor his memory.

https://doi.org/10.1515/9783110674156-202

our primary target audience, with 2020 vintage concepts and 2020s components that have extended the life of fluid machinery at best-in-class user facilities.

To extend equipment life or its operating time between interventions of any kind is the primary goal of this book. There are important side benefits to equipment life extension in the form of enhanced equipment safety and community goodwill. We will accomplish much if we specify, design, manufacture, and install fluid machines with forethought and expertise. Conversely, we will put everything at risk if we remain indifferent and are happy with the status quo.

Existing machines present opportunities for valuable upgrading with the components described in this text. Retrofitting can be done during scheduled maintenance or while engaged in an unscheduled repair event. Involved are improved components, better lubricant application provisions, and other ways to safely extend operating life. It follows that implementing experience-based procedures and insistence on suitable upgrades will yield increased profitability.

However, before moving forward with upgrading one's vintage process plant machinery, it is important to take a step back and assess the situation at one's own or, in case the reader is employed by an Engineering and Procurement Contractor (EPC), an appreciative client's process plant. Although it is true that upgrading a facility's machinery will, by definition, contribute to improved equipment availability, being aware of the full picture will prove valuable. The next chapter, Chapter 1, makes that point convincingly.

References

[1] Bloch, Heinz P. "Improving Machinery Reliability," 3rd Ed., (1998), Gulf Publishing Company, Houston, Texas.

[2] Bloch, Heinz P. and Claire Soares. "Process Plant Machinery," 2nd Edition (1998) Butterworth-Heinemann Publishing, Oxford, UK and Woburn, MA, ISBN 0-7506-7081-9.

[3] Bloch, Heinz P. "A Practical Guide to Compressor Technology," 2nd Edition (2006), Wiley Interscience, Hoboken, NJ, ISBN 0-471-72793-8.

[4] Bloch, Heinz P. "Optimized Equipment Lubrication, Oil Mist Technology, and Storage Preservation," (2020) Reliabilityweb, Ft. Myers, FL, ISBN 978-1-941872-98-7.

Contents

Chapter 1
Why extending fluid machine life is still an essential pursuit

1.1 Upgrading vintage process machinery: the full picture

Ideally, the upgrading of pumps, compressors, steam turbines, and other fluid machines should start with improved component design. Improved components must be specified by the purchaser and the specification input must come from informed people. These are informed technical people whom we will identify as Reliability Professionals. There are also junior managers who play an important role, and there are, ultimately, senior managers whose approval must be sought.

Reliability Professionals act based on knowledge. Looking at accurate failure statistics points out components that had to be replaced or repaired more frequently than others. Quite obviously, these should be upgraded before they again find their way into a modern plant. In other words, these "weak links" should be designed-out before they find their way into new machines. However, for the most part, Reliability Professionals concern themselves with upgrading components in machines that have been in service for decades but have experienced above-average maintenance frequencies, cost outlays, or even repeat failure. Later in this book, we will discuss upgrades that, once made, were of immediate benefit. Some of these had perhaps been hidden from the eyes of the non-expert or were simply overshadowed by other priorities or experiences. Collectively, however, these upgrade events represent lessons learned. As these lessons were absorbed and evaluated, they showed how critically important machinery should be specified, designed, constructed, installed, commissioned, operated, and maintained.

The big payoff will materialize in specifying the *next* machine with upgraded components. A Reliability Professional must make the business case for the incremental cost of specifying, designing, fabricating, installing, and maintaining improved machines or improved parts. These professionals must teach, brief, and enlist junior managers who will seek the backing of senior managers or corporate heads without whose approval the brakes would be put on the figurative wheels of industry.

In the year 2020, refineries, petrochemical facilities, and other process plants are increasingly complex and equipment-intensive; some of these incorporate more than 100,000 assets. Suppose we are in the business of managing a processing plant, a department within the plant, or we just want to make a difference as individual technical persons, as Reliability Professionals. This clearly implies that we understand the importance of delivering value to our stakeholders for our own good – meaning our sense of self-worth, self-preservation, and for the good of our company. We

https://doi.org/10.1515/9783110674156-001

would quickly realize that upgrading one critical piece of equipment may not always improve the bottom line if another critical piece of equipment in series with the one that was upgraded also has poor availability. Peeling away one layer at a time may be the best that we can do, but maximizing our use of the best available computerized technology is important. Doing so will greatly facilitate the process of managing data and information when time is of the essence. Indeed, time usually plays a hugely important role in delivering results. As we well know, time is money.

1.2 The path forward

The road ahead requires that we will have identified and will have made plans on what needs to be done and then to establish a priority listing. The listing becomes our schedule for going forward; our decisions on upgrading and sustaining equipment performance must have a goal. For lack of a better collective term, we call the goal "Operational Excellence" (or OE). Operational excellence might be used by others as a new and sweeping initiative, but for us it's whatever maximizes long-term profits and safety. That said, it should be no surprise that each of our many upgrade decisions has as its singular and consistent objective the cost-justified maximizing of equipment availability. To be available for production, one must minimize the need for maintenance and eliminate or at least reduce, the risk of random failure.

We keep in mind that best-in-class facilities report on-stream factors of 95%. In many facilities, loss of production for just one single day can mean more than $2 million in lost profits. High rates of production equate to low unit cost of production, which explains the quest for optimizing equipment reliability. Our examples will make the point. See beyond them and, whenever possible, try to embrace the underlying principles.

1.3 Turning visions into actions

Turning visions into action requires knowledge. We cannot acquire the needed knowledge by listening to anecdotes, or by over-emphasizing on-the job training. We must read, we must acquire the ability to ask intelligent questions, and to act on facts, not opinions. Achieving high equipment reliability implies that we understand the condition of our equipment, the damage mechanisms and modes of failure, and the risk of continued operation without regard to early manifestations of deviations from normal. We must learn the details of applying the proper level of maintenance and scheduling of such maintenance.

Optimizing the expense in maintaining equipment requires the judicious stocking of spare parts. Sometimes the spare better be an upgraded part, a pre-engineered improved component, ready to be installed at the next opportunity. Therefore, spending our maintenance dollars in a way that drives the reliability and availability of equipment requires forethought. In other words, the next spare part should be an upgraded version which *we* designed, or which *we* specified to be different from the part that failed. We must view every maintenance intervention as an opportunity to upgrade if upgrading is both feasible and cost-justified. We justify the upgrade by estimating the enhanced safety of people and equipment. We realize that safety is directly influenced by equipment reliability, which is totally dependent on design, installation details, conscientious maintenance, and as-intended operation of the equipment. If we gain added safety in any of these areas, we will have improved the facility. It should be easier to make the business case for improved safety.

As we move toward meeting our upgrade objectives, we must spell out practices that translate our vision into action. The focus will remain on achieving operational excellence – creating a safe and profitable workplace. We will have to transform company cultures and mold individuals into organizations that support these objectives. Starting at the top levels of a company, there must be an active commitment toward achieving excellence. Our managers must be visionaries and leaders in fostering a supportive culture across all departments and having an organizational structure that can execute the work that needs to be done. And we must sometimes nudge them in that direction ....

To make a vision materialize, we must maintain a work force that is highly trained, has the right tools, and is both skilled and motivated to achieve the desired results. Because the soul of a company is its people, the notion that one can always hire contractors or consultants is deeply flawed. Therefore, our hiring and retention practices must attract the best people. These people must have mentoring from both within and outside the company. They must be highly trained technically toward becoming the future leaders of the company and must develop the right skills in using the tools required in maintaining the facility. They must be given responsibility and be held accountable for results. Good performance must be recognized and rewarded to provide the motivation for continual success.

It follows that we always need to know the condition of our equipment. The figurative pinch points where allowing deviations from normal or anticipated conditions will cause unacceptable failure risk must be identified. Having the software tools to collect and analyze relevant equipment information and data is important. With these tools, one can gain or capture the knowledge needed for making smart business decisions; these are the decisions for upgrading and suitably maintaining one's equipment. Computer technology has enabled industry to manage large amounts of data and information in support of the above objectives. But proving the cost-effectiveness of using the best available technology for design and upgrading of processing equipment will be needed. Reliability Professionals cannot wait for disasters to jolt them into action. Their main charter and roles are to describe what needs to be done to steer clear of accumulating deviations at a figurative pinch point. Deviations at pinch points will almost certainly lead to performance deficiencies and distress. Certain types of distress become disasters.

Being proactive is one of the overarching roles of Reliability Professionals. As they proactively seek out, cost-justify, and implement steps toward improving fluid machines, they will not simply go with the lowest bidder for parts, machines, contract services, and the like. Instead, these professionals support management by identifying the best assets or – in the case of contract service providers – the most qualified ones and let them demonstrate how, in the past, these contract service providers have gained the approval of their key clients. Never fall for claims that everything is highly proprietary, and nothing can ever be disclosed. Make "Trust but Verify" your motto. Seek out facts and advise management why not to make commitments based on mere opinions.

1.4 Goals

Without goals, our plant can be likened to a sailing ship without a harbor. For a ship without a harbor, all winds are favorable winds. If, on the other hand, the ship has a destination harbor, its progress toward arriving at that harbor, the clear goal of the voyage, can be measured. It's no different with asset performance. Once we have established a goal based on Best-in-Class performance, uptime, profits, safety performance, and a host of other measurable, trackable, reportable common-sense goals, we are well on the road to Operational Excellence.

While we do not want to disagree with the saying: "You can't manage what you can't measure" we nevertheless know that not all metrics presently used as key performance indicators (KPIs) make sense. Giving the reader a listing of KPIs of questionable value is of little use, but our text will highlight several KPIs that *do* make eminent sense. Among these is a listing of machines and availabilities of fluid machines with failure frequencies (possibly) in excess of one's own averages, or greater in number than experienced by our competitors.

As we proceed to list elementary observations, consider the following situations:

- On-stream factor is about 85% (low) and profits are marginal, at best
- Operator errors are not only having a negative impact on maintenance expenses, but are affecting production and revenue
- Equipment breakdowns are frequent, i.e., a large number of Emergency and Priority 1 work orders are observed
- Predominantly reactive rather than proactive scheduled maintenance work
- Job planning is at a low level
- Repairs are not always correctly carried out the first time
- Scheduled outages are over budget and exceed the scheduled duration
- Critical spare parts are not always in stock to minimize equipment down time

- The maintenance budget is out of control and is escalating at a double-digit rate per year
- Equipment maintenance strategies are informal, vague, not optimized or non-existent
- Well-experienced people are leaving the company at an alarming rate
- Use of contractors for maintenance is not optimized and no in-house expert has jurisdiction over the contractor's quality of workmanship
- Completed capital projects are not reliable
- "Lost Time" and "Near Miss" safety statistics are trending upward

Ask now if, to some degree, this sounds like the situation in your plant or facility. Are our remedial actions and pursuits consistent with our objectives? Are our facility practices moving in the right direction? We would wish to develop a plan that ultimately drives plant profits and ensures the safety of people and equipment. The knowledge that unbiased SMEs have gained during decades of work in the industry deserves to be presented and illustrated with case studies that are time- tested. With that in mind, the depth of information in this book is not limited to only one person's experience, although the author's body of work will nevertheless provide a framework that readers can use and adapt to their situation. This text is not intended to restate what competent others have published, nor to imply that one person's journey is the only path to achieving specific objectives.

The important point to convey is that the book is far more statement of factual and time-tested experience than futuristic vision or consultant-conceived generalities. It's not a recounting of the passionate pursuit of a career-conscious dream, but rather a deliberate choice of enhancing one's sense of self-worth by adding value rather than enhancing one's career by playing games of office politics. The term "passion for work" has no place in this. The author was breathing a sigh of relief when reading Professor Calvin Newport's hypothesis and premise [1] that our work product should be such that, in case we would quit our jobs or depart for other reasons, our no longer being there would be noticed. Our work should aim at adding value, but not excel in degrees of passion for the job that would create undue anxiety, hurt our families, or cause us to die prematurely. The author's view of overtime is very simple: Apart from the occasional (and hopefully very rare) event where our unscheduled late evening presence at our place of employment makes sense, we should not be the persistent overtime workers. Having to do overtime work on a consistent basis is attributable to one of only two possibilities: The company is overloading a worker beyond what is reasonable, or a worker is so lacking in proficiency that he or she must put in more hours to keep up with the output of average employees.

Accept that mistakes will be made in executing your plan and learn from your successes and failures. Persevere when the going gets difficult because the road is long, winding, challenging and treacherous. Realize that you are not alone in this pursuit. You have access to many fine books, publications, and conference proceedings.

Stay in touch with mentors, consultants, and industry professionals willing to share their experience and knowledge. Competent manufacturers will explain their products and services, so make them your technology and training providers and let them assist you. Learn from them if they serve you well and send them packing if they cannot make reasonable contributions to business performance and business ethics! A manufacturer or pseudo-contributor lacking in these essentials will put you and your business on a dead-ended track.

What we have learned

- Instead of making work your passion, be so good they can't ignore you
- Eliminating weak links makes economic sense
- Make competent manufacturers your technology resource and training provider
- Don't waste time listening to consultant-conceived generalities
- Never fall for buying from someone who claims that their products and/or reference information are highly confidential or proprietary, and that nothing can ever be disclosed

---

Reference

[1] Newport, Calvin. "So good they can't ignore you," (2012), Grand Central Publishing, New York, NY.

Chapter 2
The people and the business case

After reading Chapter 1, hopefully, the reader will have taken time to assess the situation and write down what may be lacking or what could be improved in their organization, plant, or facility. As we examine several equipment upgrading and improving experiences, let's agree that many of the what, why, when, and how questions and answers deserve to be shared.

2.1 Everything starts with people

It has been said that *people* are "the soul of the company," so let's begin here. We assume that senior management's commitment and active participation are present in the enterprise. If not, and if the push for excellence starts from a lower job function, then the challenge for lower-level leaders or Reliability Professionals is to first secure the full backing of senior management. Management commitment must be in place for the various upgrading initiatives to take root and be sustainable. If the initiative is nothing but the "flavor of the month," it will fizzle out quickly.

We might say that without management commitment, upgrade-related steps or implementation routines will die from lack of support. Contributors who believe in the correctness of improvement will have ascertained and presented a strong business case. If management follow-up can be equated to benign neglect, their conduct violates the business essentials we have learned to call the three C's: Communication, Cooperation, and Consideration. Where the three C's are counted of little value, value-adding employees may see an uncertain future for themselves and their place of employment. They may leave the company for greener pastures and join entities that will not stifle their sense of self-worth. Observant employees laboring to keep the plant running can usually see behind the scenes and make realistic assessments of situations. Of course, observant Reliability Professionals must do their part of the two-sided commitment. They must learn good communication skills and perhaps re-state a business case in a language that managers can understand.

2.1.1 Good leaders

At best-in-class organizations, good leaders exist at all functional levels. Regardless of the layer or level where reliability engineers are placed in an organization, they must be eager to prepare a business case for upgrading by clearly articulating both problem and solution. To be accepted by management, the business case is best made in the form of recent specific examples and by explaining their impact on the

https://doi.org/10.1515/9783110674156-002

plant in terms of previously lost revenue and compromised safety. These, of course, are then compared to realistic expectations of future gains.

At the next level, good leaders will listen to people – Reliability Professionals and others – laying out issues and solutions. Going a step further, good leaders will look below the surface of what they hear to get the big picture. Before making decisions, they will seek input from others who have a stake in addressing the problem. The important thing is that good leaders will make decisions and will support you if you are persistent in making the case. Lack of action by senior management to support junior or future leaders will impose prolonged and unnecessary high stress on lower-level managers who, indeed, are among the key employees. Being unduly stressed will cause a considerate professional to eventually burn out and leave. Lack of action by upper management personnel sends the message that they, the top managers, don't care about a corporation's most valuable asset – its value-adding people.

2.1.2 You

Next, look at your organizational structure in the context of determining what to upgrade on your process machines. To answer the question "how?" you may have to obtain input from your technology providers, or from colleagues you have met at technical conferences. We call that networking. And don't overestimate the Internet. The Internet is a series of dots, and only experience tells us how to connect the dots. In the same vein, don't underestimate good technical texts. The non-readers among lower management will soon drop out. They will have difficulty understanding the technical rationale employed by Reliability Professionals. The non-readers among young managers risk endorsing or acting on someone's opinion, limited or out-of-context anecdotal recollections, hearsay, opinions, or the junior individual's personal feelings and untested intuition.

Ask if your organization has the technical people to support both the more immediate detailed implementation requirements and the long-range pursuits of keeping the plant running. Long-range actions still demand keeping our eyes on what needs to be done to make steady progress towards improving equipment reliability, safety, and availability. It matters if the number of assets in your facility is low or high. You must look at equipment complexity and the potential consequences of equipment failures.

The age and condition of your equipment must be assessed; understandably, the positions, layers, and numbers of people required to upgrade assets will vary. There are cultural aspects that seem well beyond our sphere of influence. Challenge this assumption in the interest of making progress.

The culture we see in our place of employment represents an integrated pattern of knowledge, beliefs, and behaviors. An organizational culture at or near the top of its peer group is well informed and generally receptive to further improvements. An organizational culture near the opposite side of the spectrum is largely indifferent.

At first glance, it cannot be budged. Everyone seems stuck in the "business as usual" mode. It will be the Reliability Professional's challenge to enlist senior management in slowly changing organizational attitudes, values, and customs. It will not be easy, but it must be done if a company wants to survive, and if we ourselves want to live with self-respect and dignity.

You may be the lower-level supervisor or manager who is compelled to make realistic assessments. The ease or complexity with which you tackle the assessment task will vary based on prior exposure, upbringing, motivation, and life experiences. Chances are you may want to start with small improvements and expand as the needs become more clearly defined. You may wish to add staff. However, because of the high expense in hiring, training, assimilating, and managing people, their suitability and fitness must be adequately justified in terms of how they contribute to the bottom line. Investment in people is analogous to how we invest in new equipment, but of much greater importance. It is not unreasonable to expect a tenfold return on the cost of maintaining people. Unfortunately, many companies make the mistake of viewing reductions in personnel as the most immediate path to higher profits. Although force reductions may be feasible or even necessary in some instances, this perceived path to profitability rarely maximizes plant profits in the long run.

Instead, our premise should be to retain or hire as many competent people as possible. As these competent individuals then add value toward achieving our goal of safely and profitably operating and improving a plant, we move our collective culture from a reactive mode to a risk-based proactive mode. We will need more innovators, read: "Fire-preventing" technical people; and fewer mere pairs of hands, read: "Firefighting" people, whose job it is to repair equipment. Reducing the numbers of field workers will become possible when they are no longer needed to repair equipment after breakdowns because of improved reliability. The cost of failures to equipment and the monetary outlays for repair crews, contractors, and lost revenue can be offset by measures that make more economic sense. As we nurture those who add value to an enterprise, word spreads and others will be motivated. Giving them the right encouragement, the tools, and modern technology support, the organization is lifted up, so to speak. Equipment reliability and the plant's profitability will increase. Paybacks approaching or exceeding 100 to 1 are not unusual as we ultimately evaluate the true value of concerted drives for operational excellence and the just profits achieved without ever compromising safety.

2.1.3 Who's who in our quest for excellence

A facility should be able to quickly activate a tactical team of Maintenance and Reliability Engineering Professionals, the "MaREP" team. Their charter and roles are to submit to Plant Management well-documented plans and cost justifications for the systematic removal of weak links in the reliability chain. It is their job to be fully informed on Best Available Technologies, components, subassemblies, control elements, machines, and devices to move the organization toward being best in class. Resourcefulness is needed to be fully informed. The members of the MaREP team must read, attend (and read the proceedings) of key technical conferences, and do considerable networking. Networking requires making experienced value-adders from top service provider organizations the MaREP team's technology resource.

In general, a MaREP team will include subject matter experts (SMEs) who often report to different managers. As these SMEs periodically meet to discuss and prioritize upgrade opportunities, they determine if and when they should actively contribute to making a compelling business case for upgrades. The SMEs making up the MaREP team are knowledgeable in:
- Rotating Equipment – Mechanical Engineers and Specialists
- Fixed Equipment – Mechanical Engineers and Inspection Group (Inspection group to include Inspectors and Specialists)
- Electrical, Instrumentation and Control Systems – Engineers and Specialists
- Welding, Corrosion, Coatings – knowledgeable Specialists, and a Metallurgical Engineer
- Process Engineering and Plant Operations

On many occasions, the above listed resources need to be accessed by the team members. If not part of this team, these supplementary resources must provide input and explanations on how the process and operation of the equipment may affect its availability. The team needs to have a higher-level management sponsor to keep up the team's momentum. The management sponsor must be able and willing to remove the occasional selfish instincts of the different supervisors and administrators to whom the members of the MaREP team normally report on a day-to-day basis.

Contractor involvement should be carefully weighed; consider contract employees to augment (but not replace) your in-house MaREP team. Augmenting or supplementing your department or team allows it to handle a cyclical workload, fill temporary technical gaps, and respond to requests if there is generally an insufficient need to retain a full-time person. It is important to note that full-time core company employees will look over the shoulders of outside service providers and ensure that the knowledge learned from contractors is retained within the company. As just one example, if a contracted individual is refilling a large lube oil reservoir on a major turbo-compressor, a designated employee is responsible for quality control. Your designated employee must verify that the various cleanness requirements associated with lubricant transfer are adhered to. Your company cannot possibly put itself in the position of wrecking equipment that is absolutely needed for production. There can be no such thing as saying: "We don't know what the contractor did; it was the contractor's job to do it right."

Accountability or ownership must ultimately reside with one of your trained employees or specialists. These are typically non-degreed people that have demonstrated superior performance and pride in their workmanship while working as technicians in the field. The knowledge and experience that technicians bring to this group can be impressive; it complements and often exceeds the skills and contributions of engineers.

A team of Reliability Engineering Specialists needs to include engineers in the disciplines of rotating, fixed, electrical, instrumentation, and control systems. These are the employees that are ensuring that a facility has:

- Mechanical Integrity Manuals for all equipment disciplines that define the elements of what needs to be done, responsibilities, and workflows to ensure compliance with equipment safety regulations, best practices, and maximizing equipment availability.
- The systems to manage the condition, reliability, and availability of equipment as well as optimizing maintenance expenditures. These include a computerized maintenance management system (CMMS) for work management and a world class asset performance management (APM) system.
- Identified the machines that fail more frequently than the average population (the “Bad Actors”) and analyze for root cause. Initiate corrective action and use the examples in this revised text for guidance. Focus on placing “Bad Actors” on a priority list because the payback for improvements will be most rapid.
- Put in place the strategies for maximizing equipment availability while optimizing the maintenance outlays. In large facilities, it is not uncommon for maintenance budgets to be in the range of $50–$100 MM per year. Proactive equipment strategies will result in greater control of the maintenance budget and the value gained in equipment Return-on-Investment (ROI) from these dollars spent.

2.1.4 Tools needed

The effectiveness with which a team of Reliability Engineering Specialists works with an Asset Management organization is of immense value for upgrading vintage process machinery! Tools are needed to be effective and to do one’s job well. Data culled from CMMS or relevant failure analysis records are among these tools, as are books on failure avoidance, computer software, and statistics.

Some CMMS are of no use to the Reliability Engineering function. To contrast just one example of uselessness versus usefulness, consider process pumps. If the CMMS is populated with entries such as “bearing failed,” there are no clues as to whether the failure was initiated by someone forgetting to put lubricant in the bearing, or if hammer blows were used to mount the bearing in the local shop, or if the bearings were purchased from a substandard source. A widely used text [1] lists more than 50 reasons why bearings can fail. It contains photos of each and points out that without knowing the root cause of a failure, the replacement part will also fail. The person doing the replacing becomes a parts changer instead of a solid contributor to future failure avoidance. It is noteworthy that the MaREP team has responsibility for the CMMS "tool" to be fed meaningful data. Without such data the team's effectiveness will be crippled.

Achievable life and actual life of comparable components and machines (or systems) enter into availability comparisons. A robust CMMS for work management is where all assets are assigned a unique identifying tag number; an equipment hierarchy and asset criticality are maintained. Work requests are issued for all work and charged against the equipment tag. Blanket work orders are needed to capture costs against non-equipment expenses but should not be used for process equipment. There is, for example, little or no value to have all the maintenance dollars for piping in a particular process unit charged to a blanket work order. A piping segment, as defined on the piping and instrumentation drawing (P&ID), can serve as the unique descriptor for piping. Corrosion circuits can be identified for use in scheduling risk-based inspections (RBI). Dollar amounts spent must be related to properly defined/identified assets in a process unit or facility. Analysis of expenses requires a measure of accuracy; management needs to know where the money went. Only then can we determine how well the money was spent and how it should be allocated in the future.

2.1.5 Work requests

When work requests are issued, the originators should list their observations. To the extent possible, originators should clearly and to the best of their ability define the work to be done. For example, "Fix Pump" is not as good as "Mechanical Seal Is Leaking Product." Ideally, a cooperative operator may even go as far as to state: "Mechanical seal leaked after pump was inadvertently run for 10 minutes before external flush supply valve was opened." Relaying this sequence of events would be a time saver and instant guide to a $20 upgrade. One could simply cement an engraved tag on the start station, making it read: "Commission external flush first. Do not start pump until flush liquid is reaching mechanical seal."

2.1.6 Tracking software

The CMMS or Reliability Tracking Software assists the maintenance department's planner with issuing accurate work procedures and work sequences. Without having relevant records, it will not be possible to plan this work and to ensure that materials are on hand to perform the repair. The maintenance technician does not need to guess what the symptom was that prompted action for repair. Another important piece of information to enter on the work request is the "Method of Detection." This field should be a drop-down box selecting the best choice for why this work is being requested.

After the repair work is completed, the technician closing the work order enters time spent and charges the work order with the materials that were bought from the storeroom. Most importantly, the worker should select failure coding from a drop-down box and enter in the comments field all specific observations and non-typical repairs made. Doing so will greatly assist the analyst in determining the root cause. The few minutes that the technician spends in doing these computer input tasks are far outweighed by the ultimate time saved, and enhanced accuracy imparted to the analyst's follow-up effort.

Another important piece of information that must be recorded against the asset is the impact on lost revenue. Typically, the operations group or its MaREP member track events that caused them to miss their targeted production. Perhaps this information is maintained on a spreadsheet that shows the Product Affected, Quantity of Product Loss, and the Piece of Equipment responsible for the loss. Marketing can provide the unit value of products; we can convert this into dollars or other currencies that all of us can understand.

During our work process of identifying the bad actors, we now have the two key components of loss: Revenue in dollars and maintenance expense, also in dollars, down to the asset or component level of interest.

2.1.7 Elementary failure analysis facilitated

From properly collected data it often becomes immediately apparent which assets are causing the most pain. We interrogate the data and can readily begin to identify the root cause and develop a corrective action from two well-proven facts [1]:

First Fact: The seven possible root cause failure categories for all machines are –

1. Design Defects
2. Material Defects
3. Processing and Manufacturing Deficiencies
4. Assembly or Installation Defects
5. Off-Design or Unintended Service Conditions
6. Maintenance Deficiencies (Neglect, Procedures)
7. Improper Operation

These seven are the only meaningful cause categories; indeed, all failure causes – without exception – are attributable to one or two of these seven. As we collect data supported by physical observation and at all times harmonizing with the laws of physics, we can delete five or six of the seven cause categories and search the remaining one or two by asking questions such as: "Is this problem limited to a specific asset or is it related to a brand and model of equipment?" or "Is it reasonable that of all the bearings we purchase from vendor 'X' only the defective ones always end up being installed in asset P-106?" We might ask if it makes sense for us to use a larger bearing in a pump model that is identical to thousands of pumps from the same manufacturer and operates with the very same head/flow, speed, and fluid properties. Would we not search for what makes this pump so different? What if the pipe connecting to the pumps discharge nozzle had been pulled into alignment by brute force?

As we continue to peel away the figurative layers surrounding "the core of the onion" and get closer to the probable contributors to failure, we must exercise discipline and resist the temptation of jumping to conclusions without having assembled all the pertinent facts. Premature or opinion-based false conclusions, often labeled as "cause-jumping," lead to the wrong corrective action. All three waste valuable time and resources; premature and erroneous guesswork costs money and contributes little or nothing toward achieving our goals.

Second Fact: We must also accept a fact that experience has taught Best-in-Class Reliability Professionals [1]: Failure of a mechanical part is always attributable to one of four possible agents of machinery component and parts failure mechanisms: "FRETT" – abbreviations for force, reactive environment, temperature, and time. Whether it is the pages of this book, the ball point pen on your desk, the drinking glass in front of you, or the tire on your automobile, we chose to call these "mechanical parts." Mechanical parts can only fail due to FRETT, and we have never seen these findings refuted by people possessing common sense.

To provide valuable assistance to the analyst in addressing this dilemma, nothing short of a world class APM system is required that is integrated with a CMMS and a process monitoring system. Information cannot be effectively used without an integrated system that serves to provide "one stop shopping." The days of hard equipment files, office file cabinets, and desk drawers are no longer effective. Furthermore, with today's problem of high turnover and early retirements, it is necessary for companies to retain or access copies of the knowledge that its top performers have archived in their brains. Dozens of companies have acquired APMs configured to provide the functionality of elements that have proven useful for processing plants. The value of a robust APM system goes beyond the information, search engines, and analysis tools needed to process information required by the analyst. Also provided are data and information that are useful to:

- the maintenance technician in making repairs
- assist planning and scheduling of work with equipment inspection plans
- the operators in operating their equipment within its design parameters
- an application for developing, evaluating and optimizing equipment strategies including PMs
- ensure and show that federal and state regulatory requirements are being met for the safety of human assets (called "people") and physical assets ("equipment")

- optimize the scheduling of the appropriate level of inspections that are risk based rather than time based
- developing dashboards as a management tool for monitoring activities
- using KPIs with metric views for drilling down into the weeds of looking at production units, production sections, equipment disciplines, and assets
- measuring the effectiveness of maintenance as well as understanding trends in maintenance costs

The chapter you have just read is a roadmap to equipment data collection. Data are the key to factual analysis. In contrast, word-of-mouth anecdotes are stuck in the past, and there is no future in them. They lack context and are loaded with opinions.

In any event, our remaining chapters are dealing with the central topic of *Improving Fluid Machines*. We will give fact-based overview material which will reflect the practices of a few industry leaders; we will also relate many case studies of value to the reader.

What we have learned

- Derive value from Maintenance and Reliability Engineering Professionals, the “MaREP” team
- Not all KPIs are useful; pick those that contrast you against the competition
- Data are the key to factual analysis; acting on opinions can cost you your business
- Anecdotes are stuck in the past, and there is no future in them because they lack context
- Good reliability leaders prepare a business case for upgrading
- Successful, yet elementary, failure analysis is based on “FRETT” and the “Seven Root Cause Categories” approach
- Concentrate your effort on understanding how value can be added by upgrading certain neglected components. Realize that these component upgrades have been implemented by best-in-class companies (BiCs) many years ago. These upgrades are chiefly responsible for the profitability and reliability of the BiCs.

Reference

[1] Heinz P. Bloch and F. K. Geitner. “Machinery Failure Analysis and Troubleshooting,” 4th Ed., (2012), Elsevier Publishing, Kidlington/Oxford, UK, and Waltham, Massachusetts, USA.

Chapter 3
Better rolling element bearings are available

If it rotates, it has bearings. If it has bearings, there will usually exist the need for lubrication. Only gas bearings and magnetic bearings function without a traditional lubricant, but they are not within the scope of this book.

There are hundreds of bearing types, styles, and sizes. There are also dozens of lube application methods and a multitude of oil types, viscosities, and formulations. In 2020, mineral oils for process pumps could cost as little as $1.00/liter or, if used in the hydraulic systems of the space shuttle, much more than $100/liter. Oils for bearing lubrication must be clean, perhaps even ultra-clean [1]. Disregarding any of the needed precautions or requirements is likely to have serious consequences, and Fig. 3.1 attests to that fact. While this chapter of our text pertains to rolling element bearings in all kinds of machines, the author often refers to bearings in process pumps for ease of visualization.

Fig. 3.1: Failed ball bearing with smear-prone brass cage.

The stipulations found in prominent Industry Standards apply to machines operating at predominant speeds; they may exclude unusually low and unusually high speeds. The stipulations or recommendations contained in Industry Standards are also intended to facilitate spare parts warehousing and procurement, combine reasonable cost and acceptable quality, allow installation with relative ease, and give acceptable service over time. As just one of many examples, Industry Standard API 610 (the American Petroleum Institute's pump quality standard) and many similar documents are based on input from many contributors.

These contributors or their representatives, both equipment manufacturers and users, serve on subcommittees that are tasked with developing and periodically updating a specific Industry Standard. Revision cycles for these standards can range

https://doi.org/10.1515/9783110674156-003

from a low of three to a high of twelve years. It is quite common for best-in-class (BiC) users and purchasers to develop addenda to these standards and to issue these as updated requirements. The add-ons contain stipulations aimed at further improving the long-term reliability and risk-averse operation of the machines they wish to acquire.

3.1 API 610 and process pump bearings

API 610 lists back-to-back oriented 40-degree angular contact bearings with brass or bronze cages as preferred for taking up axial thrust. However, using 40-degree angular contact thrust bearing sets with brass or bronze cages will not always give optimized bearing life (Fig. 3.1). Unless the shaft peripheral speed is in the right range and proper lubricants are removing the needed amount of heat and resist being wiped off, brass and bronze bearing cages are susceptible to smearing.

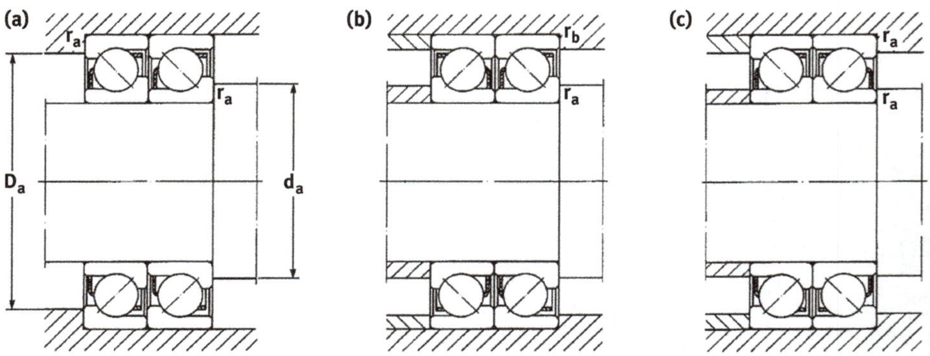

Fig. 3.2: Sets of thrust bearings with different orientations – (a) tandem, for load sharing of a pump shaft thrusting from right-to-left; (b) back-to-back, the customary orientation with thrust load on pump shafts expected in each direction; (c) face-to-face, rarely desirable in centrifugal process pumps (Source: SKF America, Lansdale, PA; also [3]).

While competent bearing manufacturers design or select contact angles that will reasonably ascertain favorable rolling motion and minimize skidding, the desirable performance characteristics of bearing sets may vary for different styles of pumps. Figure 3.2 shows but a small portion of the many different options and possibilities. For instance, sets consisting of two 15-degree or 29-degree back-to-back angular contact bearings are often best for hydraulically balanced and light-to-moderately loaded pumps operating at high speeds. As regards cage materials, the "required" copper-bearing alloys are generally more heat-tolerant than high-performance plastics, but very few, if any, process pump bearings are exposed to elevated temperatures at either assembly or during operation. The operating performance of bearings with high-performance plastics can exceed the performance of bearings with brass or bronze cages. Certain cage types may make it easier for the lubricant to reach the bearing's rolling elements, whereas other cage configurations can make it more difficult for oil to reach all parts of the rolling element bearing.

While Fig. 3.2 gives but a limited glimpse of the different orientation arrangements that are available, thrust bearing sets with contact angles that differ from each other (Fig. 3.3) will often result in favorable rolling motion and minimized skidding [2].

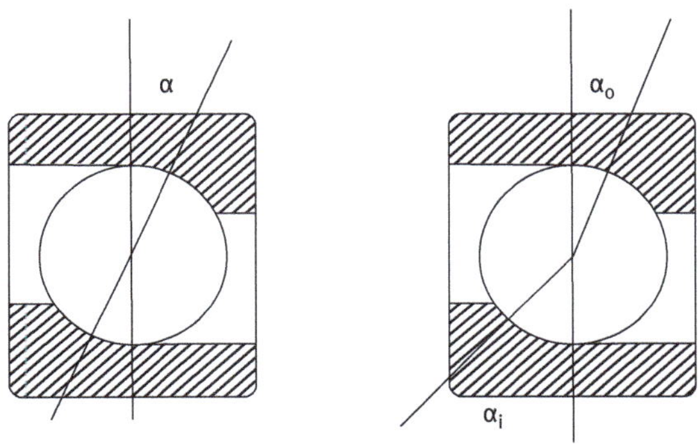

Similar contact angles, inner and outer ring. Variation in contact angles, inner and outer ring.

Fig. 3.3: Thrust bearing sets with different contact angles will result in favorable rolling motion and minimized skidding (Source: SKF USA, Lansdale, PA; also [3]).

3.2 Thrust load considerations

Process pumps with moderate thrust loads often use paired back-to-back sets of bearings; pumps with very heavy primary thrust loads sometimes use a triplex set, Fig. 3.4. In a triplex set, two 40-degree bearings are installed in tandem; the third bearing is mounted back-to-back against the bearing located in the center. The third bearing's 15-degree contour allows it to carry short-duration reverse thrust. For optimum performance, the various bearings may or may not have identical load angles.

3.3 Lubricant application choices

Lubricant application methods are of particular importance in fluid machines such as process pumps. Years of experience show that oil supplied by old-style oil rings is not always meeting the expectations of reliability-focused pump users. Special installation, design, and fabrication constraints must be observed with oil rings. Oil

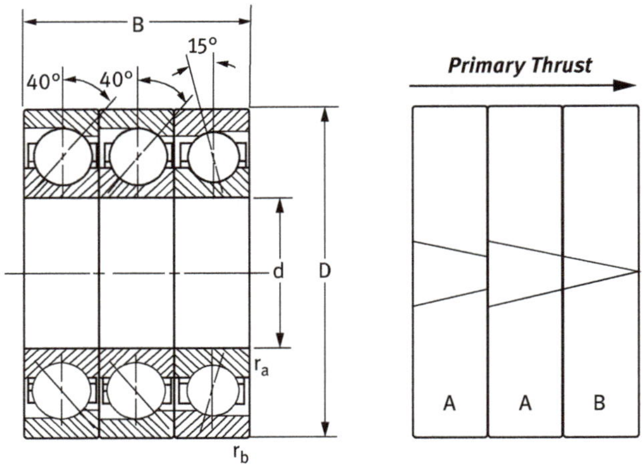

Fig. 3.4: Very heavy primary thrust loads sometimes use a triplex set. Here, two 40-degree bearings are installed in tandem, and the third bearing has a 15-degree load angle.

rings may no longer be adequate for high speed pumps in installations demanding high reliability [1, 2]. API 610 leaves the details of oil ring design to the pump vendor; however, not all vendors are aware of the issues involved, and neither are all users. We, meaning the author and his professional reliability engineering peers, will have more to say on the subject in one of the later chapters.

Whenever two or three bearings are mounted adjacent to each other, lubricant application concerns will take on greater importance. A small amount of oil applied near the front edge of the first bearing might not easily travel to the back edge of the third bearing. Likewise, a drop of oil applied at the edge of the third bearing might not readily reach, or flow toward, the first bearing. To complicate matters, certain cage styles and their respective angles of inclination will produce a fan effect, called “windage.” Depending on bearing orientation and largely influenced by the direction from which the lube oil is applied, this fan effect can oppose the direction of oil flow [1].

Again, a close review of the introductory pages to API 610 will show that the guidelines and/or recommendations of this widely used standard describe minimum requirements. It is thus inferred that the user-purchaser will often find, and sometimes even need, better or more suitable components. Reliability-focused buyers keep in mind upgrade options which can include oil mist, oil spray, and superior bearings [2]. These owner-buyers consider API 610 a commendable effort to standardize components; such components will facilitate warehousing and commonality of maintenance. Standardized bearings allow installation or assembly with only a minimum of measuring needed by the person doing the work. However, while they may be adequate and ideal 90% or even 97% of the time, standardized bearings can be a source of short MTBR (mean time between repairs) in the remaining 3% to 10%.

Some of the savings in time and money derived from standardization are beneficial, but other hoped-for savings may be quite elusive. This is exactly why the API's foreword and certain special notes in the API standards leave open the option to upgrade. In other words, knowledgeable users procure more reliable components or configurations whenever these are deemed necessary. To our point: Better bearings ***are*** available and ***are***, if appropriate, being purchased and installed by Best-in-Class performers [1]. Variations in contact angles (Fig. 3.3) serve a purpose and are available from leading manufacturers. In the late 1970s, world-scale bearing manufacturer SKF developed their PumPac™ bearing sets together with engineers at an Exxon Chemicals (now ExxonMobil) affiliate plant in Texas, USA. It was clearly shown that properly selected thrust bearing sets with dissimilar load angles can be problem solvers and are well worth their incremental cost.

Recall the triplex bearing set depicted in Fig. 3.4. Two of the three bearings are tandem-mounted and, assuming they were precision-ground by the manufacturer, will share the thrust load in the primary direction. The third bearing is mounted back-to-back against the bearing in the center. Its load angle is 15 degrees and will be activated during short duration thrust reversals only. Bearings selected with suitable preload values (Fig. 3.5) will often outperform the standardized back-to-back oriented 40-degree angular contact bearings. In fact, while standardized 40-degree angular contact bearings are generally acceptable in most process pumps, they can also be a source of elusive repeat failures.

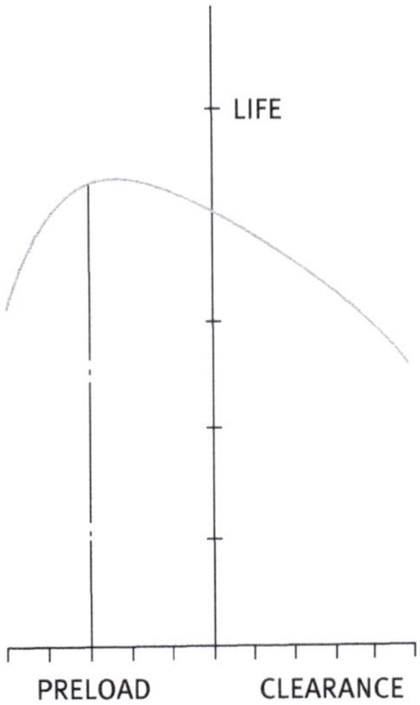

Fig. 3.5: Slight preloading prevents skidding and increases bearing life (by typically 15%). Operating with excessive preload or with bearing-internal looseness cause bearing life to decrease [3].

3.4 Competent vendors explained

Quite evidently, an interested purchaser will likely benefit from access to vendors with application engineering staff, although, sadly, the cost-cutting efforts of some bearing manufacturers have led them to discontinue retaining application engineering experts. Their websites are not disclosing the contact addresses of competent design engineers. It can be reasoned that these manufacturers want to sell bearings at the lowest possible cost, and their marketing strategy no longer includes striving to become your long-term technology resource.

An observant user-purchaser will quickly learn that it pays to be highly selective when choosing a bearing supply source. Choose one with application engineering support. Pay a premium for the products of this manufacturer. Make that manufacturer your technology resource and benefit greatly from understanding the many interacting parameters affecting pump bearing life.

3.5 Bearing internal clearances are important

One of the prerequisites to understanding pump bearings relates to the fact that bearing elements under load will deflect or "flatten"; the rolling elements will make area contact with the contoured raceways. Merely for the sake of illustration, try to visualize how, under load, the contact "point" where a one kilogram bearing ball touches the bearing race and "flattens" to an area of perhaps 0.01 square millimeter. In its simplest (and admittedly non-scientific) form, contact pressure can be envisioned as kg-per-square-millimeter. Dividing 1 by 0.01 will show the "pressure" against a bearing race as 100 kg/mm^2. If the flattened area happens to be only 0.0001, dividing 1 by 0.0001 would represent a "pressure" of 10,000 kg/mm^2. In the first case, an oil film, in compression load, will separate the steel ball from the steel race and the bearing will survive. In the second case, the bearing may not survive because the needed oil film will be thinner than the asperities (peaks and valleys visible under a microscope) and metal-to-metal contact will be made. One can deduce that the rolling elements of a survivor bearing will deflect under load, which is the only point we wish to make in this context.

Along similar lines of thought, many texts show that sets of 40-degree angular contact bearings will not be ideal for every application; neither will all types and directions of bearing-internal looseness be ideal. Bearing-internal clearances vary from a small negative amount to a particular and pre-defined internal clearance. In Fig. 3.5, a bearing is assigned a 100% design life with zero internal clearance. It can be seen from Fig. 3.5 that a small negative clearance (also called preload) will result in a slightly longer bearing life, whereas a large internal looseness would reduce bearing life to slightly below 100%.

The axial internal clearances vary greatly for different size pump bearings. For example, in the nominally 60 to 70 mm bore/shaft size range offered by different bearing manufacturers, the axial internal looseness values presently range from zero to 52 micro-meters (microns). It should be noted that any of the above can work well, although each has its pluses and minuses. Moreover, important installation and operating guidelines may differ, depending on preload and/or clearance values.

Bearings with greater axial clearances have "room to expand." Used in matched (flush-ground) 40-degree pairs, they will allow installation on somewhat oversized shafts. Although expanding the bearing's inner ring diameter, this size increase will simply reduce the pre-existing internal clearance. However, once operating, bearings with large internal looseness pose a somewhat greater skid risk than bearings with zero internal clearance or no looseness. As axial thrust created by impeller hydraulics loads up one of the two bearings, the adjacent bearing will be unloaded and tend to skid (Fig. 3.6). Of course, skidding generates heat and rapid surface degradation. Skidding can greatly limit bearing life.

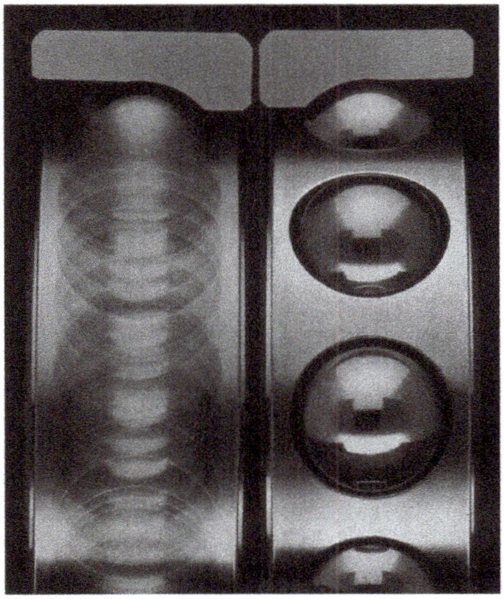

Fig. 3.6: Skidding bearing (left) vs. rolling bearing (right). Skidding generates heat and causes damage to the unloaded bearing (Source: MRC Bearing Division of SKF America).

In another scenario, if an inexperienced user-operator finds a hot bearing housing and pours water on the housing, free thermal growth of the outer ring will likely be impeded, and the internal bearing clearance will decrease. While the clearance reduction will somewhat reduce the skidding risk, cooling water does not come for free [3]. Moreover, properly designed pump bearing housings with rolling element bearings and a suitable synthetic lubricant will simply not require any cooling water. This fact is acknowledged in API 610 and is attested to by the uncooled bearing housings in an estimated 400,000 process pumps in hot service. Thousands of these operate quite successfully in high temperature pumping services. Actual bearing and lubricating oil temperatures are typically 10 degrees C (18 degrees F) lower without (!) a cooled bearing housing than with (!) a cooled bearing housing [4]. Partial cooling water jackets, meaning jackets that surround only 180 or 200 degrees of a bearing's 360 degree periphery, have been known to force bearing outer rings into a slightly out-of-round (oval) and obviously undesirable "squeezed" condition.

Visualize a set of two bearings precisely ground for zero axial clearance; their rolling elements in the 180-degree arc from 3 o'clock to 9 o'clock will be loaded, although the rolling elements nearest the 6 o'clock location will see the highest load. In contrast, a bearing with large internal clearance will have its entire load acting on only the rolling elements from 4 to 8 o'clock. The numerical value of a load on bearing elements is related to bearing life, as is the shaft's peripheral speed.

Bearings with zero internal clearance will have to be mounted on shafts with a closely controlled and rather low interference fit, perhaps no greater than 0.008 mm (0.0003 inches). If the shaft is made of stainless steel, its thermal expansion may be as much as 17% greater than that of a conventional ferrous alloy steel shaft. In that case, an interference fit of 0.008 mm (0.0003 inches) may be too tight. Abnormally high interference fits could expand the inner ring and cause a preload of unknown magnitude. But the zero-clearance bearing pays back by not requiring cooling water and by almost never skidding.

3.6 Changing the culture

The question is: Will a facility's maintenance work forces do their part and learn the different maintenance details that go with different bearings? Will a facility endeavor to make a truly competent vendor-manufacturer their training resource and technology provider? The answer depends on a plant's culture. Culture is defined as the way people do things. Culture is an integrated pattern of human knowledge and behavior; it is outlook, attitude, and shared value. Cultures may have to be changed or, at the very least, nudged into a different direction in order to capitalize on the opportunities shown in this text. Whether a plant's reliability professionals will prevail depends on how persuasively they will make the business case for better components. Their persuasiveness can change management attitudes. Managers are encouraged to ask for relevant training to be institutionalized and to observe closely the extent to which the entire organization supports the training efforts needed for this type of directional adjustment.

A reliability-focused company takes important first steps by grooming and supporting their reliability engineers. At the same time, reliability professionals must earn the support of their managers by reading, absorbing, and communicating relevant information. Reliability professionals must follow up by making the business case for purchasing bearings from the most qualified manufacturer. As but one of many examples, experienced reliability professionals ask bearing manufacturers to answer questions relating to the axial clearance on 7000-series angular contact pump bearings.

In all instances, reliability professionals should look for the facts and reason on the matter. They should visualize how, on a bearing with large internal clearances, the load will act only on the rolling elements located in the 5-to-7-o'clock positions. In contrast, in a zero-clearance bearing the load is more evenly distributed on the rolling elements located between 4 and 8 o'clock. It should be easy to judge in which of the two bearings the load per rolling element will be higher; the manufacturer's computerized analysis will show the actual numbers. Although seemingly of minor importance, the film strength demands on the lubricant will be favorably influenced by distributing a given load over a larger number of rolling elements. A bearing salesperson without pump application knowledge may not be able to explain why his product should be considered for a facility's pumps. That said, experienced reliability engineers strongly advocate bearings to be supplied only by vendors and manufacturers with a sales force that includes application engineers. A reliability-focused plant should cultivate good business relationships with vendors' application engineers and go well beyond making them their product supplier. Making them also one's technology provider will result in rapid payback for the incremental cost of purchasing the provider's superior products. Consistently obtaining full application engineering support and service commitment adds much value to an enterprise.

As to the mechanical end of process pumps, good engineering and thoughtful maintenance result in long pump life. Good engineering and extended trouble-free pump operation are achieved by accessing and considering all the facts relating to bearing styles, clearances, lubricant application method, and installation requirements. Understanding how components work is of great importance. Engineers and reliability professionals will be motivated to dig up and evaluate low-risk procedures; they work hard to define and implement experience-based best practices [5]. Of course, best practices are in no way limited to petrochemical plants and oil refineries. Best practices are followed by leaders in all major industries, including utilities, pharmaceutical plants, and even bulk consumer goods plants.

Chances are that the application engineering groups of competent bearing manufacturers were of great help to best-in-class industry leaders. Nevertheless, while all manufacturers are interested in making a sale, not all bearing manufacturers will be motivated to assist purchasers with expert advice. Make an all-out effort to work with those who are willing and able to give experience-based advice. As they soak up experience-based advice, the user's professionals will absorb training that moves their facility or company closer to the interacting goals of greater safety, reliability, and profitability. These points are repeatedly and authoritatively made in [2].

What we have learned

- Bearings per API 610 are acceptable most of the time, but most of the time is not as good as all the time. Using API-compliant bearings can be the reason for repeat failures.
- Flush-ground 40-degree angular contact bearings can cause prolonged headaches for the user.
- Making competent vendor-manufacturers one's technology providers justifies paying them a premium and will often result in huge returns. In many cases the payback will be surprisingly rapid.
- Strive to modify a prevailing culture of indifference, if necessary.

References

[1] Bloch, Heinz P. "Pump Wisdom: Problem Solving for Operators and Specialists," (2011), John Wiley & Sons, Hoboken, New Jersey, USA, ISBN 978-1-118-04123-9.

[2] Bloch-Geitner-Ehlert. "Optimized Equipment Lubrication, Oil Mist Technology and Storage Preservation," (2019), Reliabilityweb, Ft. Myers, Florida, USA, ISBN 978-1-941872-98-7.

[3] Bloch, H.P. and Budris, A.R. "Pump User's Handbook – Life Extension," 4th Edition (2013), Fairmont Publishing/Taylor & Francis, Lilburn, Georgia, USA, ISBN 0-88173-720-8.

[4] Bloch, Heinz P. "Improving Machinery Reliability," (Editions 1982/1988/1998), Gulf Publishing and Elsevier Publishing Companies.

[5] Bloch, H.P. "Petrochemical Machinery Insights," (2016), Elsevier Publishing Company, Oxford, UK, and Waltham, MA, ISBN 978-0-12-809272-9.

Chapter 4
Understanding value calculations using bearing housing protectors as our example

Bearings are precision components. To survive, they require clean lubricants in adequate amounts. Even seemingly small amounts of contamination can greatly reduce equipment reliability and uptime. Bearing housing protector seals are designed to keep oil in and contaminants out.

Beginning around 1880, natural rubber and elastomeric compounds began to replace cotton, flax, and other packings in machines with rolling element bearings. Lip seals of various configurations were devised and quickly recognized as bearing protector components. A 1940-vintage lip seal activated by a garter spring is shown above the simulated centerline in Fig. 4.1.

However, well-fitting lip seals tend to create grooves in the shaft surface, and ill-fitting lip seals don't seal. Once wear occurs, a leakage path exists for the oil, and inward movement of dust-laden ambient air or rainwater can take place more easily.

Fortunately, new generations of improved bearing protectors are now available that can help maintain lubricant cleanliness and prolong the life of rotating equipment. One such bearing housing protector seal is shown below the centerline in Fig. 4.1.

Figure 4.2 makes a compelling case for preventing water intrusion into bearing housings. However, making the business case will still be necessary. Many texts will point out that with this amount of water bearing life will not exceed 30%, or 12,000 hours of the typical 40,000 hours stipulated by the widely used pump standard of the American Petroleum Institute, API 610. With their typical leak-free life of only 2,000 operating hours, lip seals will not usually serve us well.

In making the business case for rotating labyrinth seals, we can either use direct data or shortcut-style, generally applicable, rules of thumb. Both will be discussed next.

4.1 Comparing cost of ownership

Making the business case for improvements requires calculating the cost of ownership and/or the cost of upgrading. In this chapter, we use bearing housing protector seals as our example and show how long it will take to obtain payback by implementing multiple upgrades.

https://doi.org/10.1515/9783110674156-004

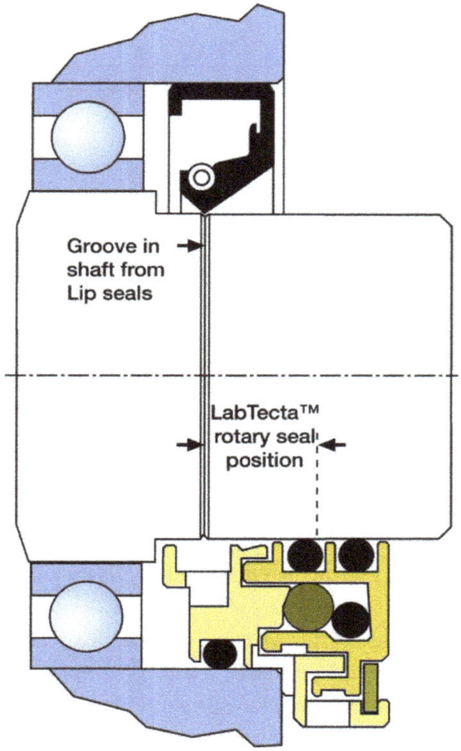

Fig. 4.1: Typical lip seal (top) and advanced rotating labyrinth seal (bottom) (Source: AESSEAL, Inc., Rotherham, UK, and Rockford, Tennessee, USA).

Fig. 4.2: Without effective bearing housing protector seals, water intrusion is likely.

4.1.1 Comparing cost of ownership

A comparison of the total cost of ownership between a lip seal and a labyrinth seal will prove revealing. Our comparison assumes a cost of $20 for two lip seals (with 4 changes per year) versus $600 for two rotating labyrinth seals (with one change every 4 years); in each case, the cost of maintenance labor would be $1,000 per event. Lip seal replacements would cost ($20 + $1,000) × 4 = $4,080 per year or $16,320 over 4 years. Rotating labyrinth seals would cost ($600 + $1,000)/4 = $400 - per year, or $1,600 over 4 years. The cost of ownership of the rotating equipment with lip seals is about ten times the cost of ownership with modern rotating labyrinth seals! By using the available O-ring replacement kit (about $60 each), the cost of ownership of the labyrinth seals can be reduced even further.

4.1.2 Calculating upgrade justification by empirical rules of thumb

Rule-of-thumb-based upgrade justification calculations use a different approach. While perhaps not as precise as the one just submitted in conjunction with the lip seal versus rotating labyrinth comparison, three rule-of-thumb calculations have proven useful in making the business case for upgrades documented in this book:

An empirical assessment, our **first rule of thumb**, assumes that a single available upgrade measure, perhaps upgrading to advanced bearing protector seals Figs. 4.3 through 4.5, will extend safe operating life by factors ranging from 1.1 to 1.4.

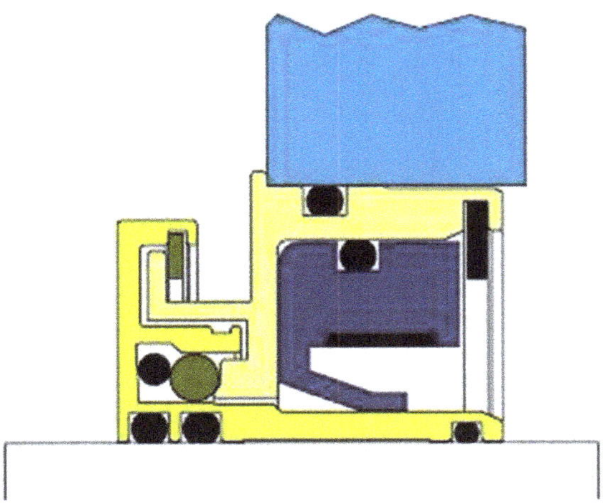

Fig. 4.3: Hybrid Lip Seals mated with advanced rotating labyrinth protector seals are designed for special applications (Source: AESSEAL, Inc., Rotherham, UK and Rockford, Tennessee, USA).

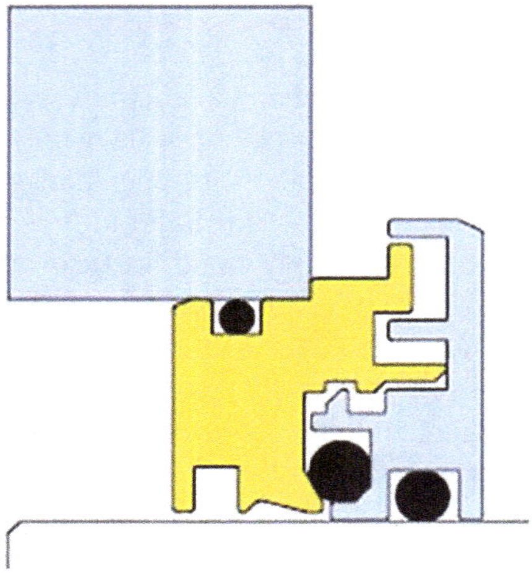

Fig. 4.4: A narrow-fit advanced rotating labyrinth seal (Source: AESSEAL, Inc., Rotherham, UK and Rockford, Tennessee, USA).

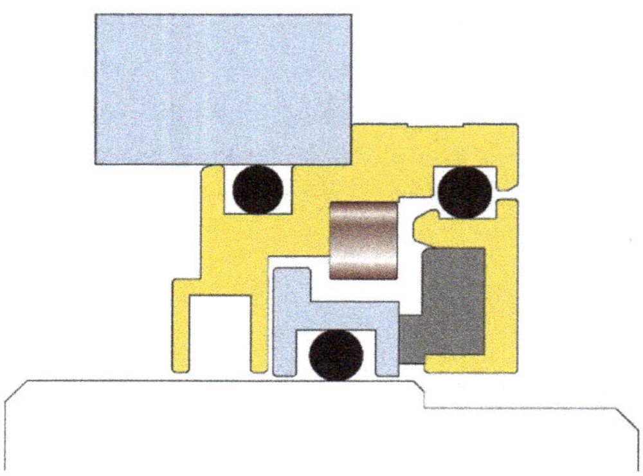

Fig. 4.5: Magnetic face seal, often used in oil mist applications. The two O-rings farthest from the shaft center line are stationary. Oil mist is introduced between an adjacent equipment bearing (not shown) and this magnetic face seal (Source: AESSEAL Inc., Rotherham, U.K., and Rockford, TN).

Implementing two available upgrade measures would extend safe operating life by factors from perhaps 1.5 to 2.5; three low-cost improvement measures get the user from 2.6 to roughly 3.3-fold operating life. These approximations are often used in initial, screening-type, cost justification calculations. They have usually yielded reasonably close results, and proceeding with upgrade plans is considered justified if payback is obtained within 18 or fewer months. In this example case, an average pump repair costs $12,000 and occurs every 4 years. We take $3,000 as the per-year repair outlay and purchase two rotating labyrinth seals for $600 to avoid one such repair. The payback is 3,000/600 = 5 times each year. That's roughly 10 weeks – not bad. We also could have said repair frequencies go from one every 4 years to one every 5.5 years. Instead of an imputed $ 3,000 per year, we now only spend $3,000/1.4 = $ 2,150 per year. We can be certain that a $600 set of two advanced bearing protector seals will last 6 years = $100 per year. We can calculate in one minute that the payback ratio is $2,150/100 or about 20:1. One-twentieth of a year equals about 3 weeks.

4.1.3 A second rule of thumb

A **second rule of thumb** uses an exponential approach. It assumes that if a fully upgraded machine has a reliability of 1.0, then a missed upgrade will lower the machine's reliability to 90% of 1.0 = 0.9; two missed upgrades lowers it to 90% of 0.9 = 0.81; three missed upgrades to 90% of 0.81 = 0.73; four missed upgrades to 90% of 0.73, equaling only 0.66, and so forth. We consider this elementary rule of thumb rather optimistic; reliability with four deficiencies is probably less than 50% of what would be achievable with better bearings, better mechanical seals, better couplings, or whatever other upgrades are achievable.

Say, we had requested the purchase of a $600 set of advanced bearing protector seals after hearing that a neighboring "Refinery X" is routinely doing this for a critical pump. We know that our critical pump has an MTBR (mean time between repairs) of 3 years and assume that theirs reaches 3/0.9 = 3.3 years. They, too, spend money on pump repairs. We spend $12,000/3 = $4,000/year, "Refinery X" spends $12,000/3.3 = $3,600/year. Over a 6-year period we spend $24,000; they spend $21,600. Their $600 upgrade returns $2,400. Their mindset puts "Refinery X" on the track to routinely do these and similar upgrades. They either are or likely will soon become a best-in-class (BiC) performer. Their market valuation probably tells the story of doing things smart and with forethought.

Then there is a **third rule of thumb** worth sharing. Again, a reasonable assumption is made; a probable 20% improvement in failure avoidance, or repair cost reduction, or life extension, is thought to results from each upgrade. In that case, an upgrade initiative will move the equipment reliability from 1.0 to 1.2, a second (different) upgrade would capture 1.2^2 = 1.44; further upgrades 1.2^3 = 1.73, and 1.2^4 = 2.07. The implementation of four proven upgrade measures would cause the MTBR (mean time between repairs) to be extended two-fold. Yearly repair expenditures would be one half of what they were before; employees previously with repairs would now spend time on repair avoidance tasks. Safety would go up, community goodwill would be given a boost, and so would worker morale.

4.1.4 Using pressure-balanced constant-level lubricators

In this example, we opt for routine upgrades by (a) using sets of advanced bearing protector seals ($600), (b) switching to extended life synthetic lubricants (incremental cost $200 per charge), (c) installing pressure-balanced constant-level lubricators, Fig. 4.6 (incremental cost of $100), and (d) a purpose-designed stress-relieved (annealed before final machining) brass oil ring ($200). The four upgrades are thought to shift the operating life of a key asset from previously 3 years to now 3 × 2.07 = 6.2 years. Suppose our records showed that it costs $18,000 to repair the process pump at issue here, and repairs would be made every 3 years. Distributing $18,000 over 3 years equals $18,000/3 = $6,000/year. A one-time expenditure of $1,100 results in spending $18,000/6.2 = 2,900/yr. Prorated savings are about $3,100 during each of the next six years. Payback time will be $1,100/ $2,900, which is less than 5 months.

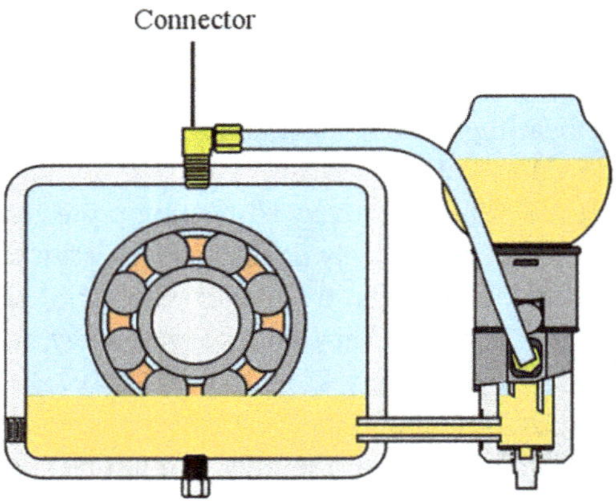

Fig. 4.6: Pressure-balanced constant-level lubricator (Source: TRICO Mfg. Co., Pewaukee, WI).

4.1.5 Where contaminants come from

Back to our bearing protector seals and an examination of where contaminants come from. Moisture and dust often enter bearing housings through old-style, ineffective, labyrinth seals, or worn lip seals. Moisture is airborne water vapor, and it could also be a stream of water from hose-down operations. Contaminants frequently enter through a breather vent, or from the widely used non-pressure-balanced constant-level lubricators depicted in just about every pump text in print today [1, 2]. An often-overlooked source of oil contamination is abraded oil ring material, which will be discussed later in this text.

Unless the rotating equipment is provided with suitable bearing housing seals, an interchange of internal and external air (called "breathing") takes place during alternating periods of operation and shutdown. Bearing housings "breathe" inasmuch as rising temperatures during operation cause gas (air) volume expansion, whereas temperature declines at nightly cool-down, or after equipment shutdown, cause the gas (air) volume to contract [2]. Open or inadequately sealed bearing housings promote this back-and-forth movement of moisture and dust-laden, contaminated ambient air.

4.1.6 How to stop the contamination

If given free communication with the ambient air (due to no bearing protector seals being used), the bearing housing will "breathe" as a function of changes in temperature. We of course know that hot air rises and moves toward a cooler environment. To stop this breathing and resulting contamination, there should be no communication or connection between the housing interior and the surrounding ambient air. Breather vents normally supplied at the top of bearing housings should be discarded, and the threaded port fitted with the connector is shown in Fig. 4.6. Note that the traditional open-to-surrounding-air constant-level lubricator has been upgraded to the balanced lubricator shown in this image.

Instead of the widely used (non-pressure-balanced) constant-level lubricator, which allows dirty air to contact the oil, a pressure-compensated (or "balanced") constant-level lubricator (Fig. 4.6) should be installed [3]. Surprisingly, even the pressure-balanced lubricator in Fig. 4.6 suffers from a seldom recognized water intrusion path: The caulking that bonds the oil-filled glass bulb to its support casting. Sooner or later, caulking will develop tiny fissures or cracks, especially when exposed to greatly differing temperatures in the course of the years. Rain or hose-down water will run down the glass bulb and find its way into the oil through capillary action. Unless a superior bonding method is found, constant-level lubricators should be replaced every five years. This makes a compelling case in favor of oil mist, as discussed in a later chapter and in even greater detail in [4].

Finally, the best possible sealing of the bearing housing requires the use of face seals. An API 610-compliant magnetically activated face seal which is effectively used on oil mist-lubricated bearing housings was shown in Fig. 4.4; a more widely used rotating labyrinth seal (Fig. 4.5) is used to seal conventional oil-splash lubricated bearings. The use of a face seal, along with the other recommendations above (such as deleting the breather vent and using balanced constant-level lubricators), will prevent the entry of virtually all external contamination into the housing. However, none of these measures will avoid contamination from inadequate oil rings and potential defects introduced by one of the old-style bearing housing protector seals shown in Fig. 4.7 [5]. Advanced bearing housing protector seals will be re-visited in Chapter 5.

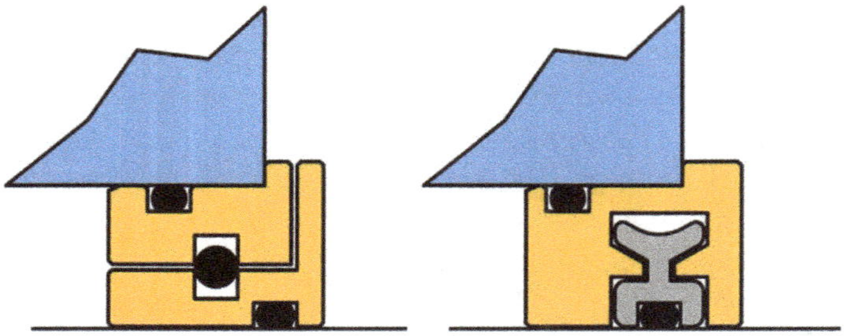

Fig. 4.7: Old style bearing protector seals with (on left) sharp-cornered O-ring semi-grooves that could cause the surface of this "Flying O-Ring" to degrade. On the right, a wedge-shaped elastomeric rotor tends to drag and seize (if too tight) or allow in-and-out leakage, if too loose.

What we have learned

- Advanced rotating bearing protector seals have largely replaced lip seals in modern general-purpose fluid machines.
- The cost of ownership of modern rotating bearing housing protector seals is highly favorable when compared to elastomeric lip seals.
- Pressure-balanced constant-level lubricators excel over their non-balanced precursor models.
- Excluding contaminants from lubricating oils pays significant dividends.

References

[1] Bloch, H.P. "Pump Wisdom: Problem Solving for Operators and Specialists," (2011), John Wiley & Sons, Hoboken, New Jersey.

[2] Bloch, H.P. and Budris, A.R. "Pump User's Handbook – Life Extension," 4th Edition (2013), Fairmont Publishing/Taylor & Francis, Lilburn, Georgia, ISBN 0-88173-720-8.

[3] Bloch, H.P., "Petrochemical Machinery Insights," (2016), Elsevier Publishing Company, Oxford, UK, and Waltham, MA, ISBN 978-0-12-809272-9.

[4] Bloch-Geitner-Ehlert. "Optimized Equipment Lubrication, Oil Mist Technology and Storage Preservation," (2020), Reliabilityweb.com, Ft. Myers, Florida, ISBN 978-1-941872-98-7.

[5] Bloch, H.P. and Chris Rehmann. "Understanding Bearing Protector Seals"; UPTIME Magazine, Issue June 2009.

Chapter 5
Upgrading general purpose steam turbines

Ever since the late 1800s, small steam turbines have incorporated bushing-type single-piece and/or segmented multi-piece carbon gland inserts (Fig. 5.1) to contain or throttle escaping steam. Simple deflectors or flinger discs were attached between the steam gland and adjacent bearing(s) in the expectation that steam leakage along the shaft would be greatly reduced. However, true leakage prevention has never been achieved because turbine shafts are made from steel, and the steel's coefficient of thermal expansion differs considerably from that of carbon. As a small amount of steam then rushes through seemingly insignificant gaps at sonic velocity, an erosive effect known as "steam cutting" causes the gap to widen. Physics, thermodynamics, and hydraulics combine to explain the interacting processes that lead to leakage.

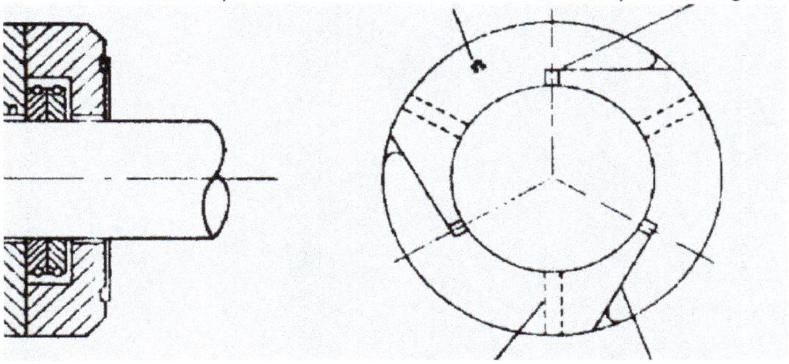

Fig. 5.1: Partial steam gland (left) and segmented carbon rings (right).

5.1 Consider two elements of bearing protection for small steam turbines

5.1.1 Optimizing both steam gland selection and bearing housing protection

In most small machines, there is then a need to limit steam egress through optimized steam gland selection. Best-in-class (BiC) facilities have long recognized that both contaminant ingress and oil leakage can be curtailed by using advanced technology bearing housing protector seals. Although inexpensive, the predecessor lip seals are no longer used by reliability-focused plants for sealing at the bearing

https://doi.org/10.1515/9783110674156-005

housing. It has been established that lip seals typically last only about 2,000 operating hours, which is a mere three months. Also, when lip seals are too tight, they cause shaft wear. In some cases, lip seal and O-ring degradation cause lubricant discoloration, known as “black oil.” Once lip seals have worn and can no longer seal tightly, steam, air, oil, and other fluids are lost through leakage.

The inadequacies of lip seals are recognized by the API 610 (American Petroleum Institute) standard for process pumps, which disallows lip seals and calls for either rotating labyrinth-style or contacting face seals. Rotating labyrinth-style and contacting face seals are taking the place of lip seals; both were shown earlier in Figs. 4.1 and 4.3.

Fig. 5.2: Steam leakage is wasteful and impairs reliable operation if steam reaches bearings.

5.1.2 Drawbacks of segmented carbon gland inserts

Small steam turbines, also known as General Purpose Steam Turbines, often suffer from steam leakage at both drive and governor-end sealing glands (Fig. 5.2). Whenever these glands incorporate segmented carbon rings (Fig. 5.1) located in a housing (the “gland”), these segmented rings are prone to leak for the reasons given earlier. Reduced leakage is obtained from allowing time to run-in and observing timed start-stop-cool cycles that are linked to temperatures reached during successive cycles. The underlying science was mentioned above: Carbon segments and steel shafts have different coefficients of expansion.

Unless timed start-stop-cool down cycles are observed during the typically eight-hour duration run-in cycle, the leakage gap will grow due to progressive erosion. So long as equipment owners take the time to monitor and implement proper run-in, segmented carbon rings may serve them well. However, if a facility does not follow procedures, or decides to entrust maintenance tasks to workers who overlook or disregard these run-in requirements, repairs will increase.

In the interest of reliability improvement and life extension, reliability-focused equipment owners give serious consideration to implementing the better solution described in this chapter. First, however, we would do well to become familiar with applicable terminology. Also, we should keep in mind that upgrading the steam turbine shown in Fig. 5.3 would include replacing both constant level lubricators in Fig. 5.3 with the balanced version shown earlier in Chapter 4. The reader may revert to Chapter 4 for more detailed information.

Fig. 5.3: Small steam turbine with bearing regions "O" located next to steam glands that are intended to prevent steam from leaking into bearings.

Each of the two bearing housings in Fig. 5.3 is located adjacent to one of the two steam glands; the ones shown in Fig. 5.3 contain four carbon rings. As mentioned earlier, as soon as these segmented carbon rings begin to wear, high-pressure and high-velocity leakage steam finds its way into the adjacent bearing housings. Traditional labyrinth seals have proven ineffective in many such cases and only solidly engineered seal glands incorporating either Dry Gas Seal (DGS) or advanced high-temperature mechanical seal technology succeed in blocking the passage of leakage steam. However, for seal faces using DGS technology, the steam must be dry and clean. If pure steam is not available, seal face combinations using advanced mechanical seal know-how are preferred over face technology derived from DGS experience.

The feasibility and cost-effectiveness of using bellows seals instead of segmented carbon inserts in small steam turbines was established and reported in 1985 [1].

As of about 2017, best-in-class facilities have discontinued the use of segmented carbon rings. The old-style steam glands shown in Fig. 5.4 (note one gland adjacent to each turbine wheel) would be replaced with the steam gland cartridge illustrated in Fig. 5.5. Different views of advantageous steam gland cartridges are shown in Figs. 5.7 and 5.8.

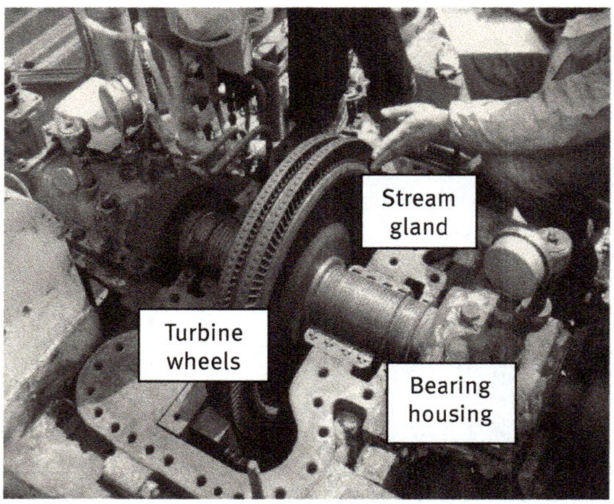

Fig. 5.4: Major components of small steam turbines.

Fig. 5.5: Steam Turbine Gland Cartridge using Dry Gas Seal (DGS) technology replaces the steam gland shown in Fig. 5.4 (Source: AESSEAL, Inc., Rotherham, UK, and Rockford, Tennessee, USA).

5.2 Bearing housing protector seals are the second line of defense

Returning to the bearing housing protector seal in Fig. 5.6, its configuration and material selection were specifically designed for steam turbines. This protector seal incorporates a small and a large diameter dynamic O-ring. Both static and rotating components in this protector seal assembly are highly stable; the rotating component is not likely to wobble on the shaft, and the entire seal is field-repairable. At normal shaft rotational speeds, the smaller of the rotating ("dynamic") O-rings is flung outward and away from the larger O-ring. The larger cross-section O-ring is then free to move axially, and a micro-gap opens.

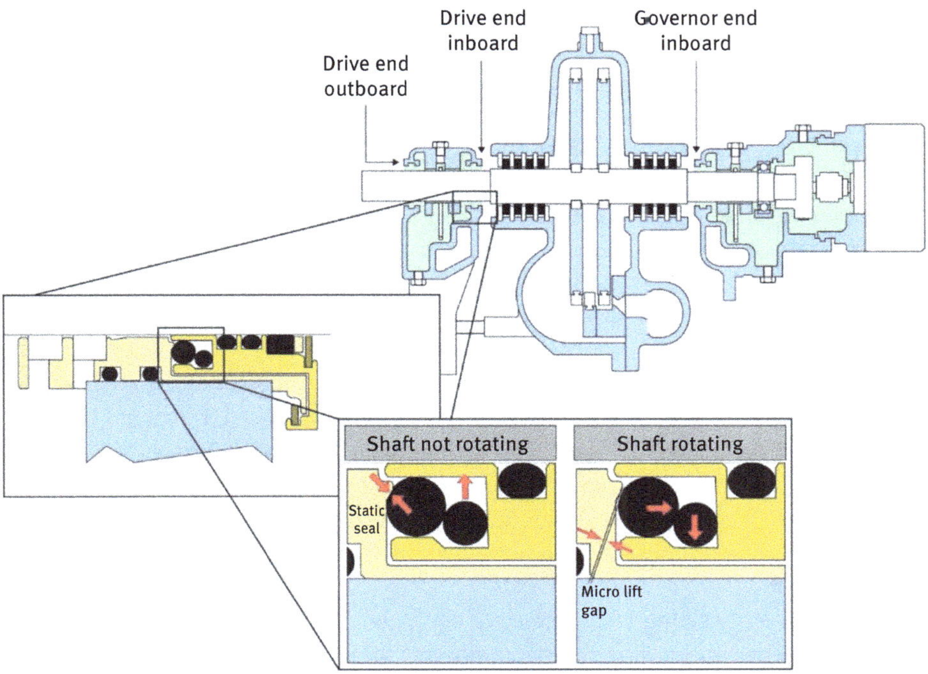

Fig. 5.6: Bearing protector seal designed for steam turbines (Source: AESSEAL Inc., Rotherham, UK and Rockford, Tennessee, USA).

When the turbine is stopped, the outer of the two dynamic O-rings in Fig. 5.6 will move back to its stand-still location. At stand-still, the outer O-ring contracts and pushes the larger cross-section O-ring. In this highly purposeful design, the larger cross-section O-ring then touches a relatively large contoured area. Because contact pressure = force per unit area, a good design will aim for low pressure. In this protector seal design, the pressure is low because a large, well-contoured sealing area is present. It should be noted that good designs differ greatly from the many technologically outdated configurations wherein contact with the sharp edges of an O-ring groove risks causing O-ring damage.

The need to keep airborne dust away from bearings is considerable in plants that process powdery substances. Although external air supplies are not normally needed, these and many other conceivable applications benefit from an engineered variant. Clean and dry instrument-grade air introduced into one of the small chambers of the bearing housing protector seal made the air bleed variant a highly effective extender of lubricant and bearing life.

5.2.1 Planned upgrade implementation speed up work

Experience shows how concerns as to the time it might take to upgrade to advanced bearing housing protector seals have been fully addressed and overcome. In June 2009, a major oil refinery in Holland arranged for the installation of the bearing protector seal shown in Fig. 5.6 in one of its small 350 kW/3,000 rpm steam turbines. The owner's facility allowed no modifications to its existing equipment, and installation of three of these protector seals on the first machine had to take place during a plant shutdown scheduled for the same month.

With no detailed drawings of the bearing housings available, the exact installation geometry could only be finalized after dismantling the small steam turbine. One of the main problems was the short outboard length: Less than 0.25 inch (~6.3 mm) was available due to the presence of steam deflectors and oil flingers. But the bearing protector manufacturer's engineers were able to modify the advanced design in Fig. 5.6 to fit into the existing groove of the OEM (Original Equipment Manufacturer's) labyrinth seals. Delivery was made within one week of taking measurements of the steam turbine. The steam turbine has now been running flawlessly for years, and decades of similar success stories attest to the advantage of using advanced bearing protectors in fluid machines.

Our point is that highly cost-effective equipment upgrades are possible at hundreds of refineries and process plants. However, superior bearing protector products for use in steam turbines must be engineered for the high-temperature application and not just replicate components built for a similar purpose. The type described here is designed for steam turbines; it incorporates important advantages compared to the original equipment manufacturers' (OEM's) standard products that were originally designed for pumps only. To summarize: Well-engineered bearing housing protector seals:

- Are suitable for high temperatures
- Incorporate high-temperature tolerant Aflas® O-rings as the standard elastomer
- Provide extra axial clearance to accommodate thermal expansion
- Incorporate high-temperature graphite gaskets in the design, if desired

Whenever both the cartridge steam gland and bearing housing protector seal upgrade options are implemented, the beneficial results will become obvious. Water intrusion into the bearing housings of small steam turbines is no longer an issue at reliability-focused hydrocarbon processing and other facilities.

5.2.2 Considering component upgrade versus machine replacement option

Scrap versus repair and recondition is a major consideration in asset management. The decision can only be made by informed staffers who investigate details and make the business case one situation at a time. *Upgrading and/or replacing* decisions can overlap, and both should be considered in the context of this book. Try to work with a repair outfit that is eager to not just repair but also upgrade so as the defer (or even eliminate) the need for repairs in the future. If it makes economic sense to consider replacing an existing machine with a new one, be sure the new machine will embody the steam gland and bearing housing design upgrades that reduce failure risk. Become familiar with the characteristics and performance expectations of potential replacement machines. Think how replacement would be cost-justified and what reliability-improving features would have to be included in a sound technical specification.

As but one relevant example, become familiar with old-style segmented carbon rings in small steam turbines and give attention to the time it takes to maintain machines with such carbon rings. With these old-style parts, manpower needs to be allocated. Skills must be both present and retained. Training will be needed. Then also think of a multitude of potential upgrade opportunities on these and other fluid machines installed in your plant or facility.

Have realistic expectations. If you seek the advice of a competent repair or rebuild shop, they can only offer you upgrades with which they are familiar. If you inquire from a vendor-manufacturer whose entire focus is to sell you his latest equipment, his bias or preference will understandably be toward making a sale. Therefore, involve your experienced reliability professionals in the decision-making process. Ask them to contribute in the form of questions that uncover potential vulnerabilities. The vendor-manufacturer must answer these questions. Do not let vendors get away with claims that they cannot disclose the "secret" details of the parts or components or machines they wish to sell you.

That leads us to again re-state the importance of eliminating or greatly reducing the need for frequent steam turbine maintenance. Recall that an innovative manufacturer offers a highly cost-effective upgrade option that replaces the old steam turbine gland seal. The early successes in adopting high-temperature mechanical seals as upgraded glands in small steam turbines required fitting "loose" parts from component seals. Therefore, mechanical seal-based glands were often more difficult to assemble than the traditional segmented carbon glands. However, substantially improved cartridge versions are available today and many of these (e.g. Figs. 5.5, 5.7, and 5.8) can now point to years of solid operating experience. Time-consuming assembly by the user-purchaser is no longer needed because the parts are cartridge-mounted by the manufacturer and shipped ready for installation.

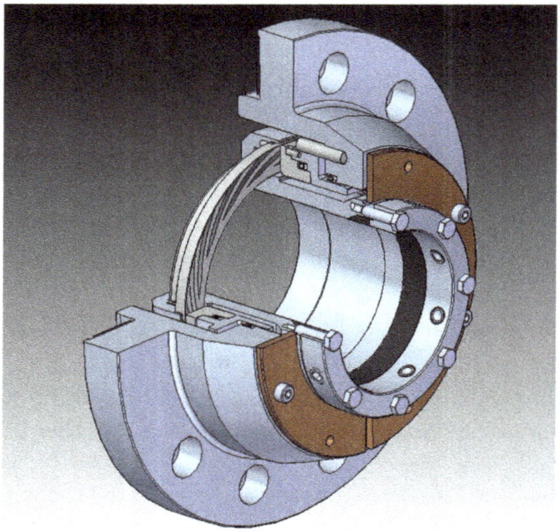

Fig. 5.7: Cutaway view of modern steam turbine gland shows its dry gas seal heritage (Source: AESSEAL, Inc., Rotherham, UK, and Rockford, TN).

As we noted before, modern high-temperature steam seals make extensive use of today's Dry-Running Compressor Gas Seal (DGS) technology. One mechanical seal provider's modular cartridge approach greatly facilitates different flange designs. Securing the rotating parts to hardened shafts is easily achieved with a clamping sleeve (Fig. 5.8) that functions like a collet in the headstock of a lathe. Moreover, highly advantageous metallurgies or face pairings are now being marketed for steam turbine gland replacement in places or at locations, where high-quality steam is unavailable.

Some products excel through simplicity of design (Fig. 5.7) and other favorable attributes. The cross-section view of the modern steam turbine gland in Fig. 5.8 shows its non-rotating flexing components sealed and centered by U-cup style high-temperature elastomers. Similar products are marketed by major mechanical seal manufacturers. There is a high probability that cartridge-style glands will be easily cost-justified once all relevant factors are assessed. A competent manufacturer or supplier will have data to prove it and will be pleased to share the data with his customers. Make this manufacturer-supplier your technology provider and training adjunct, if possible.

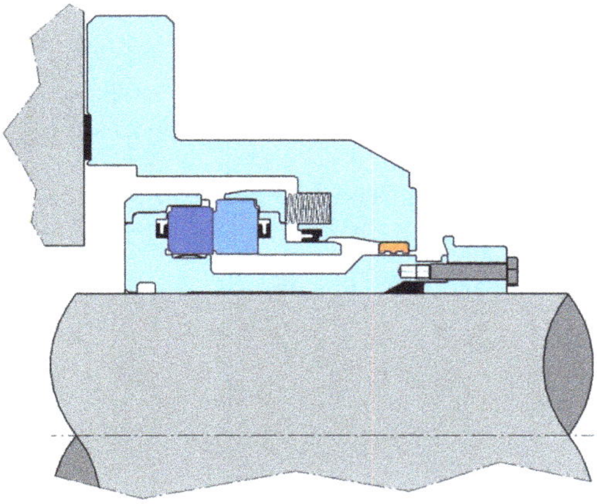

Fig. 5.8: Cross-section view of modern steam turbine gland shows its non-rotating flexing bellows components, sealed and centered by two U-cup style high-temperature elastomer parts (Source: AESSEAL, Inc., Rotherham, UK, and Rockford, TN).

5.2.3 Concluding remarks

Because Dry Gas Seal (DGS) technology has demonstrated high effectiveness in lowering maintenance costs for gas compressors, its successful adoption by other fluid machines and application in small and medium-sized steam turbines was foreseeable. Not only does the cartridge seal approach extend steam turbine life, but it will also reduce the amount of maintenance required. Equipment uptime is extended by the implementation of cartridge-style steam glands. However, the steam must be dry and clean. If steam quality is questionable, the seal face combination should be based on advanced high-temperature mechanical seal technology and experience. Either way, replacing leak-prone old-style segmented carbon glands with DGS and/ or mechanical seal-derived cartridge glands will continue to attract risk-averse and reliability-focused oil refineries and many other process plants. The same is true for bearing housing protector seals.

Finally, don't overlook the many texts that contain more insights on subjects of similar interest. Among these, consider a thorough review of [2].

What we have learned

- Steam turbine upgrading starts with high-temperature capable bearing housing protector seals
- To achieve reasonably long low-leakage life, old-style segmented carbon rings require time-consuming running-in
- Cartridge-style steam glands that are based on DGS technology are simple to install. They are effectively curtailing (or preventing) steam leakage, and no longer require lengthy run-in periods

---

References

[1] Bloch, Heinz P. and Hurl Elliott. "Mechanical Seals for Medium-Pressure Steam Turbines," presented at the ASLE 40th Annual Meeting in Las Vegas, NV, May 1985 and reprinted in Lubrication Engineering, November 1985.

[2] Bloch, Heinz P. "Petrochemical Machinery Insights," (2016), Elsevier Publishing, Oxford, UK, and Cambridge, Massachusetts, USA, ISBN 978-0-12-809272-9.

Chapter 6
Mechanical seal selection facilitated by applying principles of machinery quality assessment

The transition from the previous chapter is easy: Both Chapters 5 and 6 deal with sealing matters, and, a few exceptions aside, fluid seals are generally the weakest link in the reliability chain for fluid machines. However, they will no longer be the weakest link if seal materials, configurations, and flush plans are chosen with high-quality and long-term reliability in mind. Accordingly, we note that MQA (Machinery Quality Assessment) is successfully used to also obtain high quality, reliability, and long-term availability from mechanical seals.

Mechanical seals represent critically important components in all fluid machines. In pumps there are many different seal categories. There are well over a thousand different fluids to be sealed, and fluid temperatures range from deep cryogenic to unusually hot. To do the job most effectively, leading seal manufacturers select optimized materials and material combinations, and advocate different flush applications and flush media, propose double and tandem seal arrangements, and much more. Mechanical seal selection makes a compelling case for involving experts in the selection and upgrading processes. That said, reliability professionals should perhaps resist the temptation of becoming experts in sealing technology; instead, they should aim to be resourceful and benefit from the experience of subject matter experts (SMEs). Of course, reliability professionals must be careful not to take undue advantage of these experts by expecting them to perform lots of consulting work without compensation.

6.1 Zeroing in on best available technology

Using mechanical seals as our example, we want to next explain how one would obtain the product that best meets one's purpose. As a general rule, the single-sourcing of any critical component should be discouraged. Instead, and if possible, two to four competent suppliers or cooperative manufacturers should be invited to offer superior mechanical seals, propose the right rolling element bearings, offer only service-optimized gaskets, and so forth. Successful networking among one's peers in order to be well informed will ultimately benefit both seller and buyer. We hope to demonstrate the entire concept in this example; it deals with identifying superior mechanical seals for a difficult process pump in an existing plant.

https://doi.org/10.1515/9783110674156-006

6.2 Improving existing machine components

The reliability professional's workday is almost entirely proactive. He or she uses their available time to obtain answers to three questions:
1. Is an upgrade possible, or do we have to live with what we presently have?
2. If upgrading is feasible, is the upgrade cost-effective?
3. Which innovative technology provider should be selected to implement the combined repair and upgrade?

Needless to say: Reliability professionals work with innovators and make them their technology resource. The innovator assists reliability professionals in developing and explaining payback calculations. Together, reliability professionals and their upgrade providers research and document the prior experience. Both participate in defining the field experience at other client sites; this can be done without breaching confidentiality. The two parties often join in preparing a presentation to the reliability professional's management. At many best-in-class (BiC) companies, making such presentations is considered part of a structured MQA process, the cost of which should be included in the user-purchaser's budget. The MQA activity largely contributes to decades of future machinery reliability. MQA typically costs 5% of the asset and BiC's make sure the money for MQA is in the up-front budget [1].

6.3 MQA highlights

Details relating to MQA for large and/or important new machines can be found in Refs. [1] and [2]. As a brief example of MQA for large and important machines, we look at the super-sized integrally gear-driven compressor in Fig. 6.1. When the purchaser's representative examined the design drawings and bills-of-materials for a similar multi-stage machine at the compressor manufacturer's factory, he realized that, although specified for clean gas service, not all stages were equipped for dry gas seals (see later, Chapter 7). He negotiated a solution whereby space would be "designed in" to allow retrofit of such seals after one of the leading seal providers had accrued enough experience to qualify as a seal supplier. In other words, space was incorporated in the machine layout to accommodate an as-yet non-existent seal design during an unspecified future Inspection and Repair Shutdown (IRD) event that might present itself years from now.

It should again be noted that MQA is an activity that is well planned and for which the cost (usually 5% of the asset's price) is budgeted and allocated. At BiC companies, machines ranging from important 10 kW process pumps to huge 75 MW turbo-compressor trains are reviewed, analyzed, and critiqued by engaging in well-thought-out MQA (see also Chapter 14).

This text is deliberate in bringing MQA to the reader's attention in more than one place and on more than one occasion. Diligently carrying out this MQA activity is one of the primary reasons why BiC facilities are more profitable and reliable than their counterparts. Having an involvement in MQA activities hones a reliability professional's skills; exposure to MQA accelerates the process of their becoming value-adders.

Fig. 6.1: Large multi-stage integrally geared turbo-compressor on the factory floor (Source: MAN Turbo and Diesel, Berlin, Germany).

While originally devised for large machines about to be built, the MQA approach has also been used to evaluate existing fluid machines or the critical components that have rendered existing machines unreliable. Mechanical seal MQA was used as early as the mid-1970s when the author was asked to participate in the selection of mechanical seals for six pumping services in a large petrochemical plant. His group of reliability engineer colleagues sent out a request for a quotation to four mechanical seal manufacturers. These vendor-manufacturers were asked to fill out a simple form with 20 or so single-line questions. Among these were at least six questions and answers that were later placed on a spreadsheet for ease of comparison:

- Seal type (single or dual; flexing face rotating or non-rotating; material combination flexing face versus non-flexing face, etc.)
- Elastomer material used as secondary seal
- Pressure against which the seal operates
- Average face velocity of that particular seal
- Location where used in the fluid flow or pumping service of interest to us
- How long had seals of the type offered been in successful use at the top two locations where that exact seal was being used

Cost and delivery were of subordinated interest to the reliability engineers. Their primary focus was on deviations such as, suppose, Vendors A/B/C/D listing as their elastomers Viton (Vendors "A," "B," and "C"), and Teflon-encased Buna N (Vendor "D"). The reliability engineers then made it their business to explore if Viton was the right material or if Teflon-encased Buna N was the right material for the pumping service where mechanical seals had experienced frequent failures.

Suppose Vendor D replied: "Sorry, we made a mistake; we use Viton, like everybody else." The reliability professionals would have forgiven Vendor D; after all, everybody occasionally makes mistakes. But suppose Vendor D's answer had been: "We carefully looked at your startup conditions and realized that your fluid is gassing out for about 5 minutes. That's why we chose the considerably more expensive Teflon-encased Buna N." In that case, Vendor D would not only have made a sale but would have gained our admiration and respect. We would probably have cherished and cultivated a true professional relationship with Vendor D; we would certainly have *learned* from Vendor D.

Our take-away from this example should be three-pronged, and we hasten to point out that each bulleted item suggests effort and reward:

- Reasonable efforts to find facts will meet with substantial rewards
- Working with innovators and principled suppliers pays dividend
- Becoming best-in-class (BiC) is not a fluke; it takes work and forethought
- BiC suppliers help turning customers into BiC user companies

6.4 Select the right seal and Flush Plan

Today's oil refineries and petrochemical plants are almost exclusively using mechanical seals in their process pumps. Mechanical seals began to replace braided packing, which in the 1940s was commonly installed in the seal gland or stuffing box. But irrespective of whether deformable packing rings or seals are used, the sealing components must be good for the operating temperature and a suitable flush liquid must be introduced for cooling the shaft and its seals. However, using large amounts of flush liquid can be expensive. The flush liquid is often a refined product coming from another pump or process unit in the plant. Using Flush Plan 32 and after the clean flush liquid has done its cooling and lubricating duty (Fig. 6.2), it passes through the pump's throat bushing into the much less valuable pumped fluid. (The various Flush Plans can be downloaded from the web sites of all major seal manufacturers.)

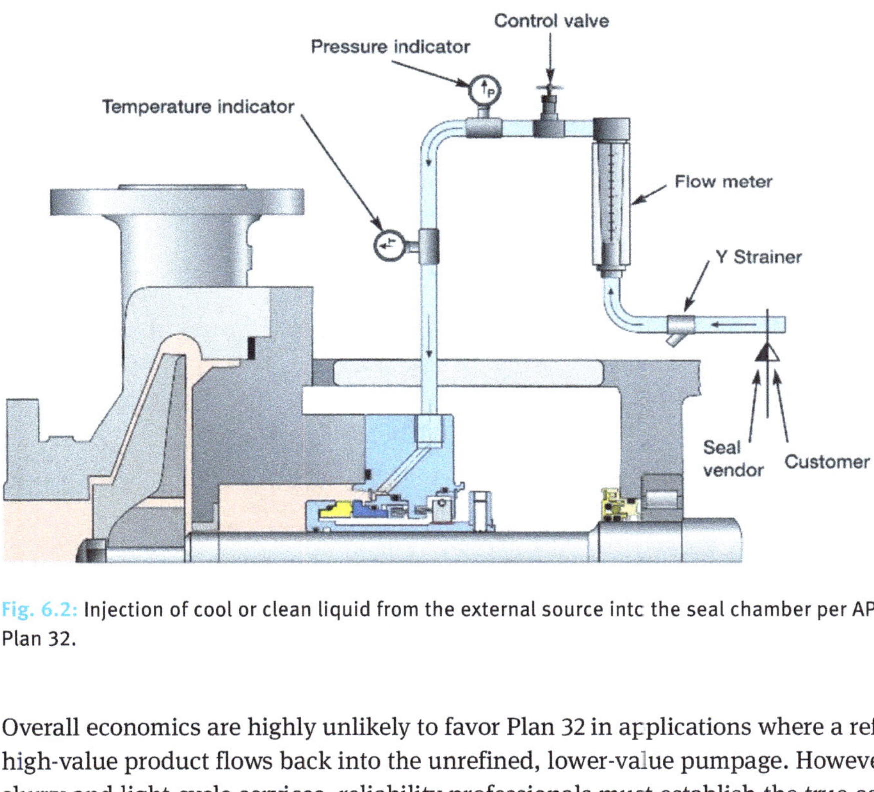

Fig. 6.2: Injection of cool or clean liquid from the external source intc the seal chamber per API Plan 32.

Overall economics are highly unlikely to favor Plan 32 in applications where a refined high-value product flows back into the unrefined, lower-value pumpage. However, in slurry and light cycle services, reliability professionals must establish the true cost of Plan 32 regardless of its simplicity, low initial cost, and overall popularity. In light cycle oil pumps that use Plan 32, relatively large quantities of clean externally-supplied flush liquid end up in the light slurry being pumped. In an e-mail exchange with the author, the former manager of a Caribbean oil refinery recalled that due to widespread use of several presumably "inexpensive" Plan 32 flush systems at that location, approximately 1,500 bpd (barrels per day) of a refined product (gas oil) were returned into an unrefined product stream.

6.5 Alternative solutions explored by a US refinery

The former manager's feedback provided the incentive for a US oil refinery to reexamine seal selection and flush practices in its coker unit. This unit had also experienced unremarkable "borderline" seal life, and the search for optimized flush application merged into a quest for both better mechanical seals and exploring optimized flush plan alternatives. A bit of research confirmed that for temperatures much above 205 degrees C (400 degrees F) one would normally favor metal bellows seals. Below 205 degrees C (400 degrees F), pusher seals (seals that incorporate O-rings) would be quite acceptable. One of many such well-reputed seals is depicted in the cross-section view of Fig. 6.3.

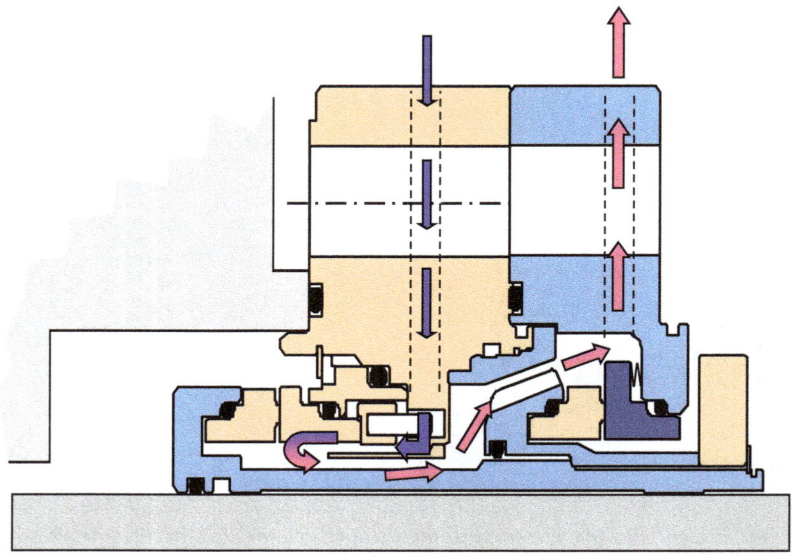

Fig. 6.3: A cartridge-style dual mechanical seal with bi-directional tapered pumping device. This seal style is suitable for many pumps in modern oil refineries in the loosely defined low-to-mid temperature range–usually not exceeding 205 degree C (~400-degree F) by >5% (Source: AESSEAL, Inc., Rotherham, UK, and Rockford, TN, USA).

For some unspecified reason, a major seal vendor had asked the refinery to consider gas-lubricated pump seals [3]. Gas-lubricated seals and "upstream pumping seals" force a clean external liquid into the seal face gap; such seals are advertised as "incorporating gas lift faces redesigned to positively pump seal flush fluid into the process."

Note that the faces of "gas lift seals" resemble those shown earlier in Fig. 5.7 for use in clean steam and shown later in a chapter on dry gas sealing (DGS) technology. But in these seals the "pumping grooves" are often only about 0.008 mm (0.0003 inches) deep. Whenever these grooves fill with solids, they will no longer function as intended. This made the refinery decide not to pursue the gas lift and upstream pumping (liquid) seal options. (It can be said that, as of 2020, these technologies are not widely applied on process pumps). Petrochemical and oil refining plants seek safe courses of action and should be encouraged to begin their search by examining the sealing practices of experienced and profitable competitors.

If we find that profitable BiC competitors use a specific mechanical seal geometry and flush plan, we would rightly endeavor to duplicate the competition's successful approach. As an aside, maintaining seal faces clean is a mandatory technical requirement for any industry segment and seal type used in process plants. Experience checks are highly appropriate before we accept somewhat generalized recommendations from marketers whose primary interests are not in line with a user's long-term reliability focus. As mentioned earlier in this chapter, the best selection approach is to cultivate access to more than just one manufacturer of mechanical seals. There are compelling reasons for having this multi-vendor access. The reasons have been widely published, and solid arguments have been made for working with two to four different seal manufacturers [2].

6.6 Best practices solutions highlighted

Following experience checks, Flush Plan 53 incorporating a heat exchanger in the flush loop was ultimately recommended for services up to about 430 degrees F (220 degrees C) at the US refinery. This choice was supported by Plan 53 being successful in comparable services elsewhere. However, it was again noted that for pumps operating with temperatures substantially above 400 degrees F (205 degrees C), the heat load in a Plan 53 circuit may exceed comfortable limits and Plan 54 (closed loop, from an external source back to the external source, not shown here) could be chosen instead. In an API Flush Plan 54, an external pump is used to circulate flush fluid in a closed loop. If maintained properly, this is the most reliable pressurized plan for dual seals. Plan 54 offers a high rate of heat dissipation and positive circulation of flush fluid.

As regards the US oil refinery and its light cycle oil pumps, at least one prominent seal manufacturer favors conventional dual seals with a higher-pressure barrier fluid. Such a seal is depicted in Fig. 6.3; it, and similar dual seals, will work well in light hydrocarbon services with moderate temperatures. The seal faces in

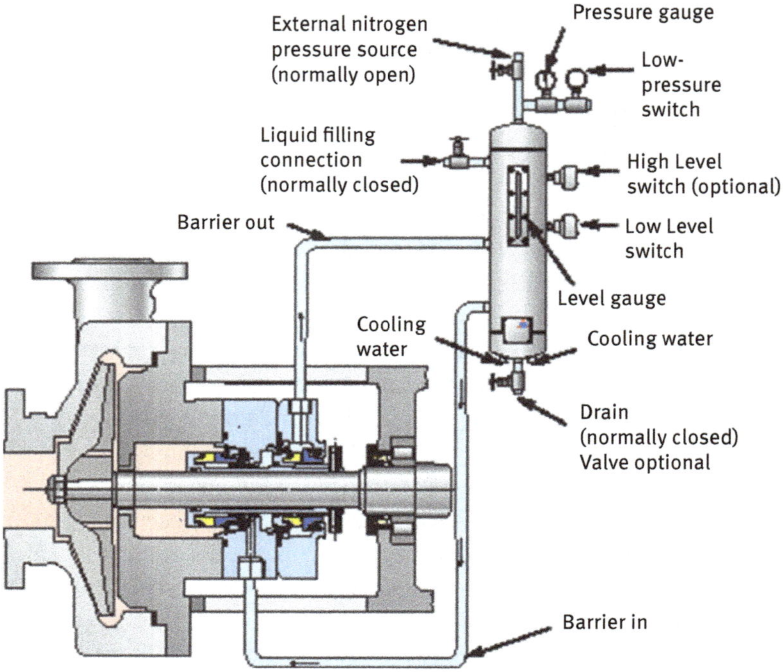

Fig. 6.4: API Plan 53A – Pressurized barrier fluid circulation in the outboard seal of a dual seal configuration. A pumping ring maintains circulation while running; thermosiphon action provides some cooling at standstill.

Fig. 6.3 are well-cooled because the seal incorporates a high-flow (bi-directional) pumping ring and a baffle that directs fluid to circulate around (and thus cool) the inboard faces. A heat exchanger is sometimes used in the flush liquid loop, although the large-diameter tapered pumping device or an auger-like pumping screw typically moves the barrier fluid at flowrates greater than those achievable with traditional configurations. Bi-directional designs generate head and flow regardless of the shaft's rotational direction.

To achieve maximum reliability, the seal design and piping plan must create the appropriate flowrate of barrier fluid to provide cooling for the seal faces. The flow must be distributed to cool both the inner and outer seal face pairs; Figs. 6.4 and 6.5 show suitable flush plans. Depending on temperature, API flush Plan 53 or one of its close derivatives is recommended and will achieve the desired effect of seal cooling and of excluding the abrasive pumpage. A very small, and generally negligible, amount of barrier fluid may still enter the pump and minimally "dilute" the product or process.

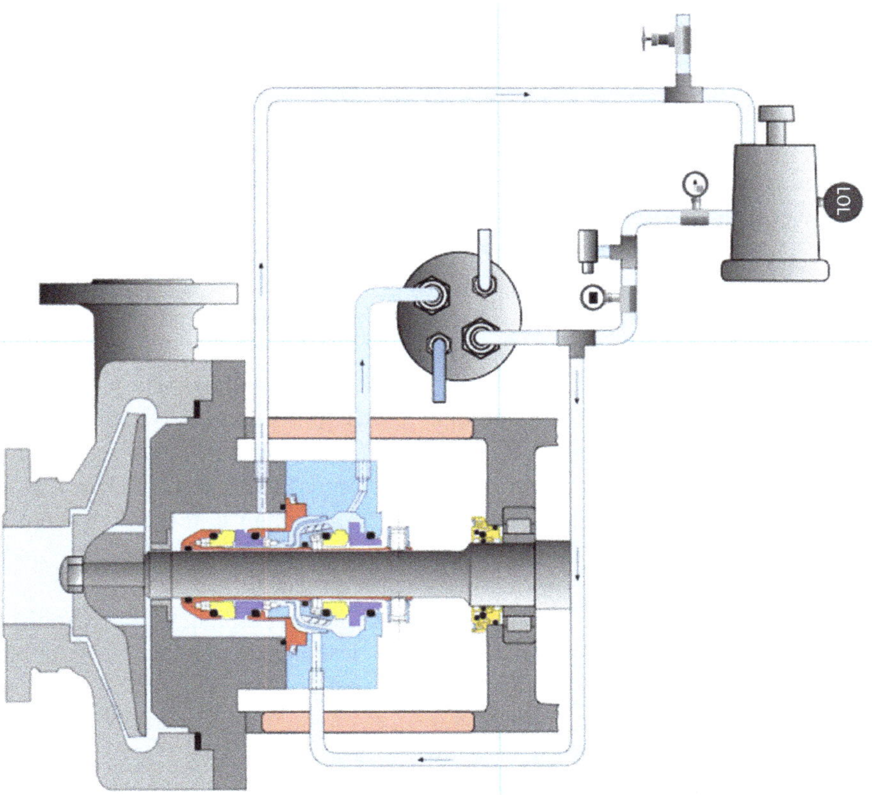

Fig. 6.5: API Plan 53C – pressurized and cooled barrier fluid circulaticn in the outboard seal of a dual seal configuration. A tapered pumping ring keeps up circulation while running.

6.7 Will the best seal fit?

A user may find that, in older pumps, the seal cavity cross-sectional clearance is probably only 3/8 inches and that some of today's advantageous seal geometries may not be accommodated by the seal cavities found in older pumps. However, several vendors offer dual mechanical seals that could fit in the spatial envelope available in the 1960s or the 1970s vintage pumps without having to modify the equipment. It should also be noted that many slurry pumps had their stuffing boxes modified decades ago after it was realized that solids tend to collect behind the throat bushing. Tapered stuffing boxes do not incorporate throat bushings; these tapered geometries and tapered bores are now widely used in applications where accumulation of solids in dead spaces must be avoided. Tapered bores can usually accommodate cartridge-style seals [3]. Other stuffing box modifications are feasible and, in most instances, easily cost-justified.

6.8 Recommendations for sealing relatively hot pumps in light hydrocarbon services

Our attempt to make generic recommendations for mechanical seals in an oil refinery is based on a case history. The recommendation was made to a refinery; it explains how, through a series of questions appropriate to the task, we will uncover relevant expertise.

We had been led to believe that trial and error solutions had been attempted by seal vendors and that the pumps in question predated 1970. Among several other possible flaws, the owners had – conceivably – overlooked that the troublesome pumps may originally have been designed for braided packing. Some of these old-style pumps were known to have rather slender shafts and braided packing acted as a "stabilizing bushing" for these shafts. This stabilizing effect may have been lost when packing was replaced by mechanical seals in the 1980s. Because of this specific possibility, suitable potential remedies were included in the list of recommendations given below. While these obviously pertained to the refinery that had initiated our discourse, the twelve recommendations are of an "evergreen" nature and will be of interest to the improvement-focused reader:

1. Establish formal communication with the four largest mechanical seal manufacturing companies (in 2020 and listed in alphabetical order, these four were: AESSEAL Inc., John Crane, Eagle-Burgmann, and Flowserve)
2. Provide e-mails to the local representatives of each, with invitation to reply by a deadline date (typically one week to ten days)
3. Provide API data sheet(s) to each of the four potential manufacturers or vendors
4. Request experience data wherein each manufacturer discloses proposed seal type and estimated or known average seal life attained
5. Make (04), above, a training experience whereby (highly undesirable) experimentation is disclosed as such. Accordingly, request feedback from the vendor who floated the idea of gas-lubricated and/or upstream pumping seals for the application at issue here
6. Request comments on any sealing system design elements (e.g., throat bushings that stabilize the shaft, "excluder" components) that keep away solids, possible seal chamber modifications required for the application [4].
7. Provide a proposal drawing with a detailed bill of materials showing the basic seal design and sealing system requirements
   a. Request each manufacturer's comment on minimum required seal cavity dimensions
   b. If warranted, consider involving seal manufacturers in your overall pump upgrade strategies or plans. In case a seal manufacturer is unresponsive or limits their interest to selling seals, consider involving a CPRS (competent pump repair shop) with applicable experience in pump upgrading
   c. Emphasize that you seek feedback on possible stuffing box modifications

8. Compare seal leakage with vendor's proposed flush plan against buffer fluid lost into the pumpage (possibly a contractor/consultant calculation) to determine incremental savings
9. Request API Flush Plan Number and/or modification recommendations to an existing flush plan. Be sure to become familiar with API Plans 53a/53b/53c and Plan 54
10. Ascertain the heat removal rate (Kcal or BTU/hr) if a cooler/heat exchanger is needed in high- temperature services. Note that some pumping ring designs are considerably more efficient than others; establish facts and vendor experience, but realize that Plan 54 is superior to Plan 53
11. Compare vendor-supplied data and look for deviations from standard, or customary answers
12. Follow up on deviations and resolve differences between the answers submitted by different vendors

The above recommendations and sequences have been validated by many decades of field experience. Granted, it might annoy some seal manufacturers to have to compile responses to the user's quest for authoritative, factual answers. Still, following these recommendations will rarely disappoint the reliability-focused seal purchaser and competent seal manufacturer. The products from two to four cooperative vendor-manufacturers will ultimately be used by the reliability professional's plant. Professional ethics will not allow procurement from less qualified or unresponsive manufacturers whose only advantage is initial pricing.

What we have learned

- BiC users see many advantages in working with more than just one mechanical seal manufacturer
- Asking questions and comparing answers will allow us to find negative as well as positive deviations from customary practices, including superior component materials and preferred flush plans
- Dual seal reliability has progressed to where they are always preferred by BiC users for reasons of safety and reliability
- MQA methods can be highly effective for identifying critical components, such as mechanical seals

References

[1] Bloch, H.P. and Geitner, F.K. "Compressors: How to Achieve High Reliability and Availability," (2012), McGraw-Hill Publishing, New York, NY, ISBN 978-0-07-177287-7.

[2] Bloch, H.P. and Geitner, F.K. "Maximizing Machinery Uptime," (2006), Elsevier Publishing, Oxford, UK, and Cambridge, Massachusetts, USA. ISBN-13: 978-0-7506-7725-7.

[3] Bloch, Heinz P. "Petrochemical Machinery Insights," (2016), Elsevier Publishing, Oxford, UK, and Cambridge, MA, ISBN 978-0-1 2-809272-9.

[4] Bloch, H.P. "Pump Wisdom: Problem Solving for Operators and Specialists," (2011), John Wiley & Sons, Hoboken, NJ. ISBN 978-1-118-04123-9.

Chapter 7
Dry gas seals: success, OEMs versus non-OEMs, questions, and what to emphasize

When it comes to dry gas seal (DGS) topics, the author is indebted to Dr. Chris Carmody, a leading expert on dry gas sealing technology. As the Special Products Manager, Compressor Sealing Design and Testing, for a large and highly respected manufacturer in the UK (United Kingdom), he is thoroughly familiar with the sealing requirements of centrifugal ("dynamic") and rotary ("positive displacement") process gas compressors. These machines require sealing components located between their pressurized gas-containing volumes and the lubricated sliding (as opposed to a rolling element) bearing assemblies, which support compressor rotors in size ranges from approximately 500 to 50,000 kW and higher.

7.1 Conventional "wet" seals

Whenever we look at conventional process compressor seals, we immediately see that lubricating oil serves as the fluid that separates rotating from stationary sealing elements. The barrier oil for these "wet" compressor seals is typically introduced at pressures approximately 2 bar (~30 psi) higher than the opposing compressed process gas. Wet compressor seals require an oil supply system that typically includes an oil reservoir, two or more oil pumps, filters, coolers, control instrumentation, and valves. The seal oil supply system in wet mechanical seals costs approximately 10% of the total compressor cost and requires a fair amount of maintenance.

7.2 Moves toward "dry gas" seals

Already in the early 1990s, some rather successful DGS applications, Fig. 7.1, became viable sealing options. Although wet seals (oil seals) will probably remain in contention during the first three decades of the twenty-first century, cost and reliability considerations have accelerated the development of dry gas seals (DGSs). Today, in 2020, seven and eight years of uninterrupted compressor operation are no longer the exception for clean gas compression services. This partially explains why, since about 1998, dry gas seals have displaced many of the different precursor seal styles.

Another recent development began in about 2006 at a highly innovative UK-based seal manufacturing company [1]. The company has extensive experience with both the design of their own and the successful refurbishment of all kinds of DGS assemblies originally provided by other seal (or compressor) manufacturers. The

https://doi.org/10.1515/9783110674156-007

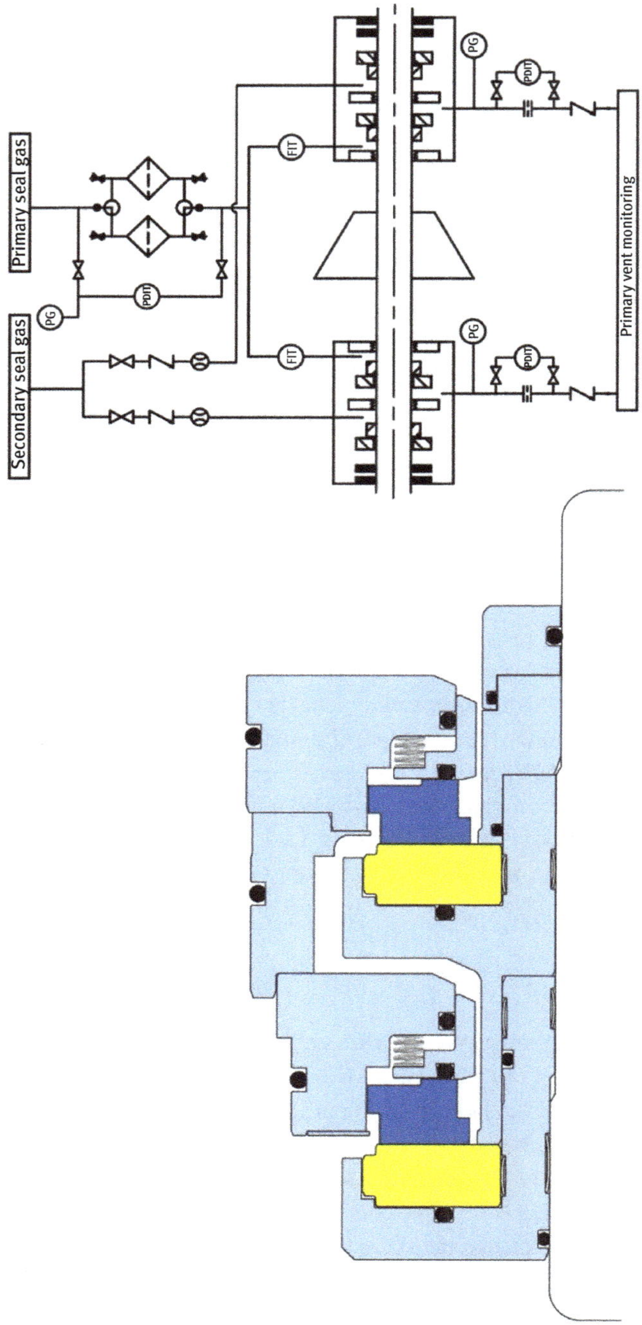

Fig. 7.1: Dry gas seal and support system (Source: AESSEAL, Inc., Rotherham, UK, and Rockford, Tennessee).

functional features of DGSs are described below; this chapter also provides an interesting cost comparison of outright seal replacement versus full refurbishment and testing.

7.3 How dry gas seals function

Simply put, DGSs operate by creating and maintaining a very thin gas film (<5 μm) between two mating disc-like surfaces, one stationary and one rotating [2]. The stationary face is spring-loaded; maintaining this gas film under all operating conditions is essential to reliable seal operation. The gas film is so thin that the most efficient method of demonstrating its existence is to perform a rigorous running test. For this reason, all DGSs should be dynamically tested, first at the time of manufacture, and then after seal repair (often called seal refurbishment). The basic schematic representation (Fig. 7.1) depicts a DGS and its associated controls. Depending on the nature of the gas and the criticality of service, the design of gas conditioning and DGS support and monitoring system can be quite sophisticated.

There are many functional similarities between DGSs and their precursor models. The precursor or predecessor oil-lubricated ("wet") seals include many variants of face, bushing, and floating ring seals. Yet, there are component features that differ in wet seals as contrasted with DGSs. For instance, the seal face of the rotating mating ring in a DGS can be divided into a grooved area at the high-pressure side and a raised ("dam") area at the low-pressure side, Fig. 7.2. The shallow grooves are often laser-etched, spark-eroded, or chemically milled. A typical depth is ~0.0003 inch = 8 micrometers (8 micron), quite obviously achieved through highly precise machining operations. T-shaped, V-shaped (bi-directional), and L-shaped (unidirectional) grooves have been produced, and each configuration has its advantages and disadvantages; some were probably devised to simply get around a competitor's patents. A stationary sliding ring is pressed axially against the mating ring by both spring forces and sealing pressure.

The sealing gap is located between the mating ring and the sliding ring. The needed gas film is achieved by the pumping action of the grooves and the throttling effect of the sealing dam. Suitably designed groove geometries are critically important for trouble-free operation of the seal. Dirt-laden grooves will not allow the seal to function properly.

7.4 Minimizing the risk of sealing problems

Upgrading from the traditional compressor seals to advanced DGSs is entirely feasible. There are cases where gas seals are a good economic choice because the user does not have to purchase an elaborate seal oil console. Suppose, however, the compressor train is a few years old and incorporates older, but technically satisfactory "wet" oil seals. Suppose also that the compressor train's already existing seal oil console has proven reliable. In such instances, it may be difficult to justify the cost of installing modern DGSs.

Table 7.1: Turbocompressor Seal Selection Guidelines [3].

Experience-based recommendations for compressor seal selection			
Application	**Service**	**Inlet pressure (1) kPa (psia)**	**Seal type**
Air compressor	Atmospheric air	Any	Labyrinth
Gas compressor	Non-corrosive Non-hazardous Non-fouling Low value	Any	Labyrinth
Gas compressor (2)	Non-corrosive or corrosive Non-hazardous or hazardous Non-fouling or fouling	69 to 172 (3) (10 to 20)	Labyrinth with injection and/or ejection using as motive gas the gas being compressed by the machine
Gas compressor	Non-corrosive Non-hazardous or hazardous Non-fouling	≥25,000 ≥(3,600)	Gas seal (4), with tandem arrangement preferred
Gas compressor	Corrosive (5) Non-hazardous or Hazardous Fouling	Any (6)	Oil seal, dual Gas seal (6)

Notes:
(1) Operating seal pressure range
(2) Where some gas loss or air induction is tolerable
(3) Pressure ranges shown for labyrinth seals are conservative. Manufacturers extend this range upward, resulting in a debit due to power losses
(4) Within a state-of-the-art
(5) H2S is the most common corrosive component in process gas compressors
(6) DGSs often have pressure limitations lower than those of oil seals.

Table 7.1 delineates experience-based application ranges and criteria used in selecting compressor seals. While these are intended as general guidelines that sometimes overlap, they are endorsed by competent and reliability-focused users [3].

Although orders of magnitude less complex and less costly than their wet seal counterparts, DGS support systems merit the full attention of reliability-focused

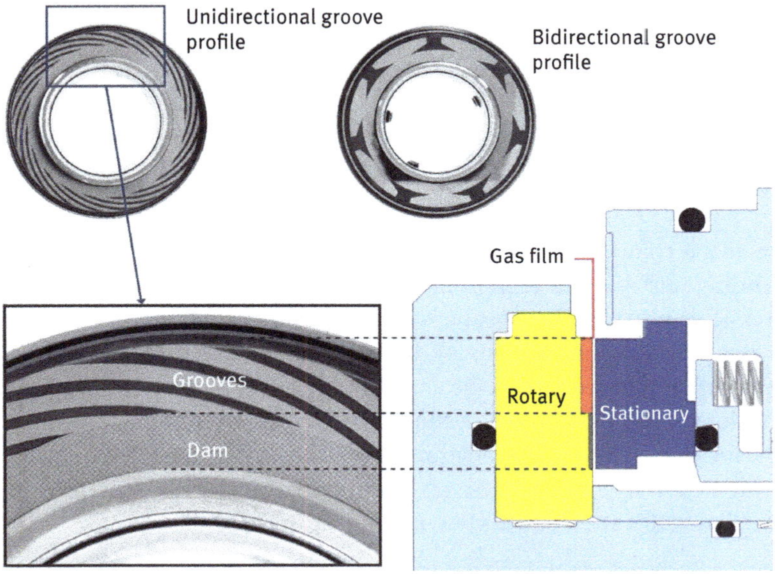

Fig. 7.2: Upper left shows face with unidirectional seal grooves intended for CCW rotation; the lower right shows it close-up. Symmetrical etchings on face depicted in upper right confirm suitability for omnidirectional rotation. Note dam common to all, as shown in the seal face assembly at bottom right [3].

buyers. DGS products deserve a measure of MQA effort. In their respective order of importance, the following factors should be considered in examining dry seal support systems for centrifugal compressors:

7.4.1 Gas composition

Understanding the actual gas composition and true operating condition is essential, yet often overlooked. For example, it is necessary to define when and where phase changes start, and that condensed liquids must not be allowed in the sealing gas. A checklist approach is often helpful:

1. Is clean and dry seal/buffer gas available at all anticipated compressor speeds?
2. Is the seal protected from the bearing lube oil?
3. How is the compressor pressurized or de-pressurized?
4. How is the machine brought up to operating speed, and how will the seal react?
5. Are all operating and maintenance personnel fully familiar with the compressor maintenance and operating manual?
6. Is the full control system included and adequately described in these write-ups?

7. Are the key elements of the system design understood, and do they include seal/buffer gas conditioning, heating, filtration, regulation (flow versus pressure), and monitoring?

7.4.2 Seal safety and reliability

Positive sealing of the compressor during emergencies must always be assured. In the event of "settling-out" gas pressure in the compressor casing exceeding that of the seal gas pressure, advanced face seal assemblies are designed with shutdown pistons. These small pistons are located and dimensioned to exert a force proportional to gas pressure. If the pressure in the compressor casing is lost or relieved for any reason, the force-action on a set of shutdown pistons keeps the seal faces closed and prevents gas release to the environment. As an interesting aside, many safety features are best tracked and observed during orderly, planned, compressor shutdowns. Function verification of shutdown pistons is one of them.

7.4.3 Non-OEMS

It's an understatement to observe why working with an experienced DGS repair service provider is always helpful. It would be fair to say that the non-OEM provider's allegiance is primarily with the seal user. Experience-based input may cast light on where DGSs may or may not provide positive gas sealing if the seal faces are damaged or distorted. The backup seal may show reasonable performance under low pressure but may fail to perform under higher pressure if the primary seal fails. Elastomeric O-rings are used in all DGSs and recall that Fig. 7.1 showed no fewer than 14 of these O-rings. They are being replaced, and seal faces re-lapped to perfection during DGS refurbishment.

The alarm and shutdown devices of the seal oil system must be of reasonable range and sensitivity to provide reliable alarm and shutdown characteristics. DGS systems often rely on pressure switches of very low range and high sensitivity. These switches might tend to malfunction or give a false sense of security. The owner-purchaser must make sure the system will provide a sufficiently high degree of alarm and shutdown performance.

Some seal configurations excel at online monitoring. The favored seal must allow easy on-stream verification of sound working condition. Beware of seal systems that are hampered by small diameter orifices that tend to plug; avoid sensitive pressure switches that often become inoperable. DGS failures are not as easy to detect as wet seal failures. In all instances the well-versed user must go through a rigorous cost justification analysis. This is where Fig. 7.3 may prove helpful; it represents an experience-based approximation of the differences between three possible options and may assist in assessing the anticipated cost.

7.5 Combining OEM design expertise and OEM repair know-how

Compressor end users, also known as owner-operators, normally face considerable expense when a machine is taken out of service for a routine overhaul. At this point, they may feel pressured to re-fit brand-new seals supplied by the original DGS manufacturer. However, given that there is usually no discount could make this "go new choice" an expensive option. Some small savings can be anticipated by having the seals repaired by the original DGS manufacturer. Still, the associated repair costs often reach up to 80% of the price of new seals – not exactly a bargain. In extreme cases, when the original DGS manufacturer has full order books and 100% test stand utilization, they may well elect to price the repair at an even higher cost than a new unit. The original seal manufacturer knows that new seals would likely affect delivery schedules. This manufacturer then justifies premium pricing because his ongoing new seal production efficiency would be disrupted by having to revert to batch size economics.

However, a more cost-effective option might be to allow a competent 3rd party DGS repair specialist to refurbish the seals. Using a typical two-seals-per-machine beam-type compressor as our example, Fig. 7.3 shows the cost comparison between the three options just outlined. A similar cost comparison should precede any "buy new" versus "refurbish" decision.

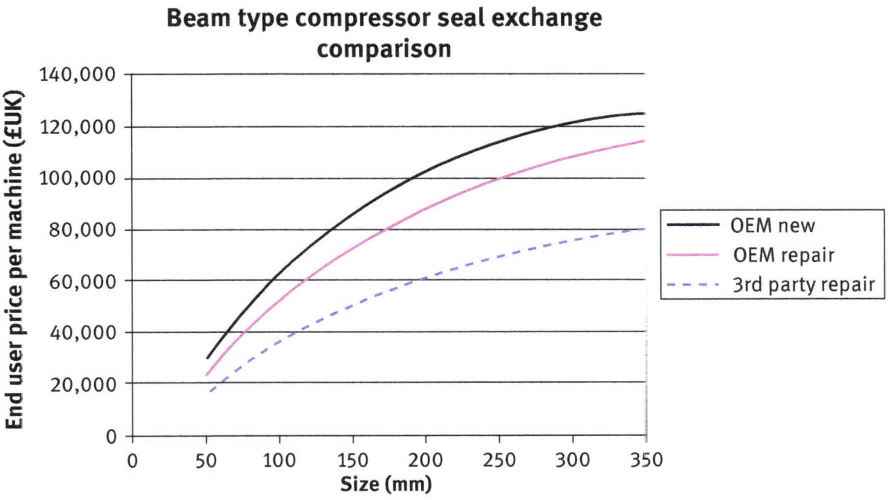

Fig. 7.3: An experience-based estimate contrasts likely cost of three different DGS exchange strategies (Source: AESSEAL, Inc., Rotherham, UK, and Rockford, Tennessee, USA).

Regardless of hydrocarbon gas prices, it will be prudent to consider gas leakage rates in order to justify conversion to gas-lubricated seals. Along these lines, one researcher [4] suggests we study the failure statistics of our oil-type seal and compare these statistics with current estimates. Current failure rate estimates for gas seals are in the vicinity of 0.16 failures per year (2020), meaning that we could expect a problem every six or seven years. At least one DGS manufacturer recommends basing maintenance intervals for DGSs on limits imposed by an elastomer's aging process. This manufacturer suggests the following maintenance routine after 60 months (or five years) of operation:

- Replace all elastomers
- Replace the springs
- Replace all seal faces and seats
- Carry out a static and dynamic test run on a test rig.

7.5.1 Emphasize seal gas systems

The author's colleague and frequent co-author (see Refs. [4, 5]) related a very costly mishap at a world scale hydrocarbon processing plant. It was reported and alleged that the incident was caused by dirty DGSs being used in their major compression train. As we then look back at our experience with compressor gas seals, we read a third colleague's words:

> Remember, at pre-startup trials of our DGS buffer gas supply system we detected some very fine iron oxide particles in the seal gas flow. This was evidenced by the fouling of the magnetic follower in the armored seal gas flow meter. Recall that we had the lines cleaned; the control valve was also removed and cleaned. It had been assumed these auxiliary lines were clean...

Our experience has shown that a continuous supply of clean and dry buffer gas is one of the most important requirements for trouble-free operation and long DGS life. Unfortunately, this key requirement is often ignored throughout the various planning, commissioning, and operating steps. This shortcoming, particularly during commissioning, is almost guaranteed to result in multiple seal failures that cause delays in plant start-up and business loss.

Seal or buffer gas filters are generally made of stainless steel with coalescing, duplex filter elements and continuous transfer valves. The filters are typically sized for a minimum of three times the normal flow, or the required flowrate for the seal failure case, whichever is greater. For a saturated or heavy condensing seal gas, consider installing filters downstream of the throttling device to maximize the removal of condensate. Consider designing seal gas filters in accordance with applicable pressure vessel codes.

7.6 Making *technically* sound decisions

Thus, our advice to the reader is to make informed choices. Consider DGSs only in conjunction with a clean gas supply. If your process gas can cause fouling deposits to develop, ask critical questions of anyone offering DGSs for use with that kind of gas. If extensive micro-filtration is needed, factor-in the cost of maintaining a DGS support system. Look for seals that will survive a reasonable amount of compressor surging. Consider DGSs that incorporate features ensuring start-up and acceleration to full operating speed without allowing the two faces to make contact. If these seals are not available from your favored supplier, look beyond the usual sources.

Recall again that at least one innovative manufacturer offers modern DGSs for original equipment (OEM) as well as aftermarket applications. Also, this manufacturer has the capability to repair and test DGSs made by others [1]. In the many years since 2009, a considerable number of users and client sites have been added to the reference list then published by our co-author's company. Finding comparable services, operating speed, etc., is an important component of selecting a vendor.

This text recommends that owner-purchasers and their Reliability Professionals investigate the extent to which any compressor or seal manufacturer can meet the buyer's reliability stipulations. The logical follow-up or conclusion would be to use DGSs from vendor-manufacturers that meet the buyer's criteria and stipulations.

7.7 Making *economically* sound choices

Most compressor OEMs have service and re-rate divisions that will upgrade, repair, refurbish, and modify anyone's machines. There are also third-party independent repairers that are often formed, owned, or staffed by former employees of compressor OEMs. That simply confirms that either the OEM or an independent non-OEM will service, overhaul, repair, and even re-rate entire compressors, regardless of origin. However, while compressor users tend to accept as standard practice any of these repair options for their compressors, the same is not true for DGSs. For reasons nobody can explain, DGSs tend to be sent back to seal OEMs. This practice deserves being labelled a paradox for the simple reason that an entire compressor train is far more complex than a DGS. Moreover, a seasoned DGS manufacturer may be eminently more qualified to do DGS refurbishments than a compressor manufacturer. Factoring in the DGS manufacturer's testing capabilities has allowed Reliability Professionals to entrust critically important DGS refurbishing tasks to companies with overlapping capabilities. These would be DGS manufacturers that not only design and produce new DGSs but also have broad knowledge of competitors' designs. Accordingly, preferred providers will have invested in facilities that refurbish and test not only their own but also other manufacturers' products on time-sensitive schedules.

DGS production for new compressors has been ramped up in the years since 2000, and so has the well-planned implementation of dry seal retrofit options. As with all industries where demand for new products outstrips the capacity to support existing products, both pricing structure and lead time for component repairs have come under pressure. In other words, primary emphasis on the manufacture of *new* DGSs has resulted in service issues related to product engineering and repair support for DGSs that have been in use for several years.

One competent DGS provider has anticipated this issue even as it began to slowly develop. In line with correctly analyzing the situation and drawing on confirming experiences gathered in the 2000–2004 timeframe, this manufacturer designed and built facilities that are specifically aimed at the repair and testing of dry gas compressor seals. As of late 2019, they had repaired many hundreds of DGS units; they are also engaging in thoroughly reviewing and updating seal gas control panels whenever such efforts are warranted.

7.8 Deliverables summarized for discussion by client and supplier

The value of testing is worth explaining again. Testing is an important feature of verifying the correctness of a repair. The selection of a DGS manufacturing entity – be it the compressor OEM, a DGS manufacturer, DGS refurbisher, or combined manufacturer-refurbisher – must be governed by experience criteria and by meeting certain testing stipulations. Unlike some OEM repair facilities that do a shortened test on repaired DGS units, select a seal manufacturer-rebuilder that fully tests all DGS units, both new and refurbished, as outlined in API 617. Only full testing harmonizes with today's emphasis on asset reliability. It has been said that modern test facilities are the key to competent repairs. Also, it is intuitively evident that properly engineered DGSs contribute greatly to overall machine reliability. Advanced seal designs reduce overall maintenance requirements; they also conserve power and lower gas consumption. They are thus contributing to sizeable operating cost savings in gas transmission services.

Some DGS test stands are limited in size and speed capabilities. Favor DGS repair service centers with expanded repair and inspection facilities that comprise two or more test rigs. Favor those with each test cells receiving power from separate substations. Ascertain that the DGS manufacturer and/or refurbishing company has capabilities that cover a full range of sizes and speeds. Each test rig should be equipped with a state-of-art variable speed drive enabling precise speed control of the motors up to a maximum speed well beyond the maximum operating speed of your compressor.

In advanced test stands, each motor provides the input to a planetary gearbox with three or more output modules that can be interchanged. Typically, the maximum anticipated speed of today's small turbomachines approaches 45,000 rpm.

Assuming your process gas compressor operates at no more than 7,000 rpm, again verify the power and speed capacity of each test cell. It should allow the dynamic testing of large diameter high speed seals under full operational conditions.

7.8.1 Test gas

Testing should be feasible with at least 350 bar (5,000 psi) air or 230 bar (3,300 psi) nitrogen. Irrespective of the type of gas used for testing, it should pass through a coalescing filtration system that removes particles above two microns in size. In order to be able to test a relatively wide range of DGS operating conditions, there should be provisions for heating or cooling.

Control of test activities must be achieved through specially designed test panels and software. The test panels should control all gas flow in and out of the test cell; they should be designed to enable any part of seals under test to be individually pressurized or vented, as appropriate. Test speed and temperature should preferably be controlled by integrated software; this software should also control the data logging or capture of speed, pressure, gas consumption, and temperature upstream and downstream of the seal while the dynamic test is in progress.

In summary: Know your requirements before linking up with a refurbishing company. Investigate beforehand who would have the needed capabilities and give extra credit to those who have the most applicable experience. Let them explain the repair process and don't hesitate to ask questions. If you are a reliability professional, your management will look to you for answers. Obtain them by sitting down with the selected provider before anyone signs a purchase order or contract. Consider including an important clause regarding exceptions to your specified requirements: "I, the refurbishment provider or person authorized to represent the XYZ DGS Refurbishment Corporation, affirm that XYZ take no exceptions to any of the client's stipulations and requirements as stated in this document."

7.9 The DGS repair process

The ability to test DGSs to high standards is only one element of a full repair service. A leading DGS service provider will likely be able to draw upon many years of experience in the conventional mechanical seal design and manufacturing industry. They will have transferred the same service-based culture into a first-class DGS repair program. The first element of the repair process is usually an initial price quotation that can also be based upon electronically transferred images or historical repair knowledge of the seal's condition. This allows the service provider to quickly respond to customer demands for reasonable estimates of cost and time-to-repair; the estimate sometimes leaves the office before a defective seal arrives at the

DGS service facility. Have the provider confirm that once a seal arrives at their facility, it will be disassembled and mapped for damage assessment. The damage should then be documented in a concise examination report that provides the full scope of work needed to restore the seal to its original as-designed condition. Firm price and delivery should be rapidly offered; the provider should also send you details of the dynamic test protocol. The test protocol should harmonize with the recommendations found in API 617 [5].

Already at this stage of the repair initial design work is conducted. This design work typically covers equipment and fixtures – such as spin test and balance mandrels, which will be needed later in the repair process. Because the DGS refurbishment company must be geared toward different seal and compressor designs, fixtures must be purposefully designed for a specific repair. Some of these fixtures will later be needed to verify the integrity of rotating face materials and, if required, to effect dynamic balancing to ISO Standard 1940. A complete material assessment must be done, whereby all materials of construction are fully certified. An experienced DGS refurbisher is expected to demonstrate a cooperative spirit by providing both rapid repair response and all requisite documentation. Ask if and how his timely response is facilitated by close links to top suppliers of raw materials. Has the DGS repair provider company made it a practice to order materials beyond those needed for a specific repair? If the answer is in the affirmative, the resulting build-up of DGS parts stock will instill confidence that scheduled deliveries can be met, regardless of other clients urgently seeking answers to their repair requests.

7.10 Replacement and rebuilding steps

Replacement of damaged faces is an important facet of the repair process. Premium grade corrosion-resistant tungsten carbide, reaction-bonded and sintered silicon carbide and silicon nitride seat replacements should be offered, as needed. Stationary faces may have to be supplied in blister-resistant carbon and diamond-like carbon (DLC)-coated silicon carbide. Where a manufacturer's experience points to advantageous PTFE (Teflon®) coatings, a reliability-focused user would ask for such coatings to be applied to seal faces.

The rebuilder should be able to repair steel components; it should be possible to recoat these as required. Where replacement parts are necessary, full code requirements should be met. In all repairs offered by competent DGS repair firms, replacing consumable items must be their default action and minimum standard. Such replacement must include all springs and fasteners; also to be included are secondary seals, where all O-rings fitted to DGS repairs are explosive decompression-resistant grades. Whenever polymer upgrading is feasible, the original selections should be replaced or superseded with spring-energized PTFE (Teflon®) or PFPE (perfluoropolyether)-equivalent sealing components.

Since most DGSs are supplied with external barrier devices, these components must also be repaired along with the seals. Labyrinth components and both contacting and non-contacting segmented seals are often in need of being supplied as part of a comprehensive repair. Providers with manufacturing capability for such parts deserve to move up in the listing of preferred suppliers.

7.10.1 Your "Specification of Deliverables"

Once all the replacement parts have been specified and produced, re-assembly and testing can proceed. An experienced DGS provider conducts full spin tests at 23% over the maximum speed rating of the seal. The component itself is thus subjected to twice its normal rotational stress. Also, whenever applicable, full dynamic balancing should be done on all rotating assemblies. Similarly, all stationary assemblies should be subjected to pressure tests that exercise (move back-and-forth) the face assembly and serve to identify seal hang-up. Only then should the various assemblies be fitted into the test equipment.

Write into your "Specification of Deliverables" that you, as the client and Reliability Professional, should be notified and invited to observe DGS testing at the provider company's facility. Even if unable to attend, clients should still participate anywhere in the world by viewing the test progress in real time by way of a suitable Internet link.

One competent provider makes available a second remote surveillance option. It involves a 3G (as of 2019) mobile phone link, wherein the entire test is being filmed with a 3G camera. This option could also be made available in the form of live video streaming to any mobile phone with suitable 3G features. Irrespective of whether you, the Reliability Professional and customer, choose to witness the tests, each test is to be fully recorded, and a CD or Flash Drive of the full test event produced for each individual seal. Ask that the CD or Flash Drive is part of the documentation (or "Information Pack") that is placed inside the shipping cartons for the repaired seals.

Although tests are normally conducted in accordance with API 617, at least one experienced DGS refurbisher often accepts and carries out additional customer testing requirements. Dynamically testing each seal typically takes three or four hours, which includes second-by-second data logging.

Client-purchasers usually ask the seal refurbisher to provide an information packet for each repaired seal; a typical data packet includes the pre-repair inspection report, a post-test condition inspection summary, full details of the dynamic test itself, a full set of tabulated dynamic test results, a graph of the data logging results, and a full set of installation details. A comprehensive report includes a spin test certificate, a balance test certificate, a dynamic test certificate, plus full certificates of conformance for any materials used. Unless an agreement requests otherwise, the repaired seal should be shipped in a heavy-duty airfreight case which incorporates carrying handles and a locking mechanism. At least one competent provider even offers an assembly tool replacement service. This is particularly useful when seal assemblies were produced at locations remote from the compressor factory, but the compressors were shipped without the best available DGS assembly tools.

7.10.2 Predefine your preferred DGS repair and reconditioning firms

It is intuitively evident that the corporate history of a seal design and manufacturing company will give clues to its overall competence. An entity that can provide seals of their own design and point to the successful refurbishment of all kinds of DGS assemblies deserves to be considered here, especially if the entity's DGS repair service continues to grow [4]. As the clients seeking to link up with a DGS service provider focus on available options, they will be impressed by a few candidates. Top choices will have received DGS models made by more than one other manufacturer. Top DGS refurbishers will have completed repairs on hundreds of these seals with a combined dollar value in the millions, and covering several OEM compressor and competing DGS product ranges.

Clearly, then, compressor owner-operators have solidly verifiable options [6]. They can decide to continue working with the OEM, or can explore, verify, and trust the demonstrated capabilities of firms whose expertise in seal design is evident. Refurbishers that impress us are obviously using modern testing and surveillance methods; moreover, they will make it easy to allow their clients to have access to these testing facilities. Good service starts with a repair facility being genuinely pleased to help asset owners, customer or purchaser desiring to be part of the entire process. Likewise, a smart reliability professional will investigate the full extent of the service provider's (or DGS manufacturer's) commitment to timely repairs of virtually every type or model of compressor DGS in use today.

What we have learned

- DGSs are mature technology and are the best choice for new machines in clean gas service
- If an existing gas compressor with a well-designed lube and seal oil system requires ("wet") seal repair or replacement, investigate if a redesigned or upgraded wet seal is available and should be considered
- DGSs should be provided with well-designed gas supply/support systems. The control panel and its associated equipment must preclude condensed gases entering the seal
- Compressor startup conditions must be evaluated by the DGS provider and provision made to supply seal gas under all foreseeable speeds and operating conditions
- DGS refurbishment should be entrusted to DGS manufacturers with demonstrated refurbishing and testing capabilities
- A detailed protocol should be specified and used for seal testing

References

[1] Carmody, Chris. "Dry Gas Seal Repair and Testing," CompressorTech2, December, 2009.

[2] Carmody, Chris, and H.P. Bloch; "Advances in Dry Gas Sealing Technology," Hydrocarbon Processing, December 2015.

[3] Bloch, Heinz P. "Consider Dry Gas Seals for Centrifugal Compressors," Hydrocarbon Processing, January 2005.

[4] Bloch, Heinz P. and Geitner, F.K. "Compressors: How to Achieve High Reliability and Availability," ISBN 978-0-07-177287-7, McGraw-Hill, New York, NY (2012), pp. 34-40.

[5] Bloch, Heinz P. and Geitner, F.K. "An Introduction to Machinery Reliability Assessment," 2nd Edition, (1994), ISBN 0-88415-172-7, Gulf Publishing Company, Houston, TX, pp. 242–247.

[6] Mid-East Turbomachinery Consulting Ltd, Dhahran, KSA; Personal reviews and professional communications with the author, 2014.

Chapter 8
Why consider perfluoro-alkoxy carbon-filled polymers

8.1 Upgrading fluid machines with advanced composites – an introduction to reliability-enhancing PFA/CF materials

Since about 2002, pump users have had remarkable success upgrading critically important process pumps with Vespel® CR-6100, a composite material made from PFA (perfluoro-alkoxy) and long carbon fibers [1]. The material is listed in American Petroleum Institute Standard API 610, 11th edition, under its generic name perfluoro-alkoxy carbon-fiber reinforced composite, generally abbreviated as PFA/CF. While PFA is chemically closely related to PTFE (commonly referred to as Teflon®), the PFA/CF composite material has superior properties that include high creep resistance, low coefficient of thermal expansion, chemical resistance, and a wide allowable operating temperature range.

Although originally used to upgrade process pumps, PFA/CF parts are no longer confined to process pumps. They have since been applied to all manner of wear components in fluid movers, and extensive references to that effect are widely available. In 2016 and 2017, the Boulden Company (Conshohocken, PA, and Ellange, Luxembourg) completed publishing an informative series of guidelines on upgrading process pumps. Boulden was instrumental in exploring the full merits of a prominent PFA/CF product manufactured by DuPont under the trade name Vespel® CR-6100. The Boulden Company's releases [1] delved into details on how to upgrade various horizontal and vertical pump types. Their releases focused on reducing the clearance of the wear rings, upgrading throttle bushings with the company's innovative PERF-Seal® design, and upgrading the line bearings surrounding vertically oriented pump shafts.

8.1.1 Consider high-pressure rings and bushings made from fiber-reinforced fluoropolymers

The now well-documented PFA/CF developments are of significant interest to reliability professionals who rightly view every repair event as an opportunity for improvement. One of the many cost-effective upgrades that can be performed at the time of repair consists of fitting composite wear rings with reduced clearance. The typical upgrade practice is to replace the stationary wear components (Fig. 8.1) with a PFA/CF composite while the rotating elements continue to be made from suitable metals. Figure 8.1 presumably shows a critically important 2.7 MW (3,600 HP), 4,500 rpm, amine feed pump. Its differential pressure is 64 bar. (928 psi).

https://doi.org/10.1515/9783110674156-008

The first plants implemented this upgrade in the mid-1980s. As of perhaps 2004, the upgrade had gained widespread acceptance across the hydrocarbon processing industry. Non-metallic wear rings are presently included in the American Petroleum Institute's Standard API 610, 11th edition (also ISO13709:2009), *Centrifugal Pumps for Petroleum, Petrochemical, and Natural Gas Industries*. As noted, Vespel® CR-6100 is listed in the current API standard as PFA/CF reinforced composite.

Fig. 8.1: A critically important 2.7 MW (3,600 HP/4,500 rpm) amine feed pump being retrofitted with Vespel® CR-6100 Perf-Seal® and impeller wear parts (Source: Boulden Company, Conshohocken, PA, and Ellange, Luxembourg).

Best-in-Class (BiC) user-purchasers make it their practice to include specification clauses that emphasize this requirement in the original machine design. As a result, the purchaser will have obtained a better pump or fluid mover from the day the equipment is put in operation.

8.2 Three benefits highlighted

Upgrading a process pump with composite material wear rings and reduced clearance provides three significant and clear advantages over metallic wear rings.

8.2.1 Avoiding pump seizures

The upgrade replaces metal-to-metal contact points in the pump with metal-to-composite contact points. Because there is no metal-to-metal contact in the pump, the risk of rotor seizure is essentially eliminated. When the composite material used is Vespel® CR-6100, the pump can even survive running dry for extended periods.

Because there is minimal risk of seizure when using composite wear parts, the clearance at the wear rings, throttle bushings, and center bushings can be reduced. This reduction in clearance improves equipment reliability and energy efficiency.

8.2.2 Improved reliability

Wear rings, throttle bushings, and center bushings create rotor stability due to a phenomenon known as the Lomakin Effect. As clearances are reduced, both the stiffness and damping of the rotor are increased. The net results are lower overall vibration levels and reduced shaft deflection – in more practical terms, longer seal and bearing life.

8.2.3 Increased efficiency

Differential pressure across the wear parts causes flow to recirculate within the pump. With composite wear parts, clearance can typically be reduced by 50% compared to metal wear parts (although there are a few fluid services where standard clearances are maintained). Reduced clearances will cut the recirculation approximately in half, potentially increasing the maximum throughput capability of the pump. Alternatively, the motor will consume less power at the same operating flowrate creating a significant reduction in the life cycle cost of the pump.

8.3 The Perf-Seal®

Horizontal multi-stage pumps offer users some of the best opportunities to upgrade with composite wear parts. Pumps delivering light or low lubricity products benefit from the non-seizing characteristics of the composite material. Long flexible shafts benefit from increased stiffness and damping from the reduced clearance. Charge pumps can potentially generate more unit throughput, leading to greater profits. Energy savings from higher efficiency in these pumps can be quite substantial. Horizontal multi-stage pumps also frequently generate high pressures. Over the years, the high pressures in these pumps have been a subject of debate concerning the suitability of composites. An innovation dating to about 2014 called the PERF-Seal® (Figs. 8.2 and 8.3) represents a significant advance in this area.

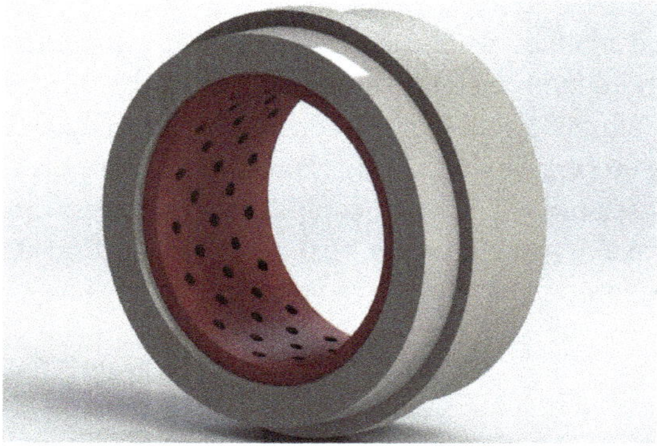

Fig. 8.2: Perf-Seal (R) (Source: Boulden Company, Conshohocken, PA, and Ellange, Luxembourg).

The PERF-Seal® is a flow-restricting part that incorporates a pattern of holes drilled through the composite bushing component in order to augment the performance of the wear part. The holes create a field of turbulence to restrict both the fluid leaking across the component and the fluid spinning within the component, creating favorable hydrodynamic effects.

The pattern of holes also increases the differential pressure limit of composite components. Previous publications of the pressure limits of composites relied upon case histories but did not address the underlying geometry of the composite component. An investigation into the pressure limits of composite parts has revealed that the failure mode under high differential pressure is due to high-pressure liquid penetrating the press fit interface at the outside diameter of the composite part. The high-pressure fluid migrating into the press fit interface creates a radial differential pressure down the length of the composite component, which can create bushing deformation, contact with rotating parts, and ultimately damage to the composite insert.

In contrast, the holes of the PERF-Seal® design eliminate potential distortions to the components due to this pressure and make the application into high-pressure pumps and other fluid movers more reliable and predictable. Current field applications (in 2019) have shown success running in differential pressures up to 170 bar (2,500 psi) differential pressure.

Finite Element Analysis (FEA) shows that this design nearly eliminates radial distortion of the bushing and dramatically increases the component's differential pressure capability. In fact, there can be additional benefits of improved throttle bushings: Testing has shown that the hole pattern discourages tangential whirl, which increases the ratio of beneficial damping to de-stabilizing cross-coupling forces. This can reduce any tendency toward damaging rotor instability as the pump running clearances wear.

The same testing program demonstrated that the holes in the I.D. (inside diameter) surface restrict axial flow across the bushing, contributing an incremental efficiency gain beyond the increase in efficiency already experienced from using PFA/CF wear parts with reduced clearances. Lomakin effect shaft support stiffness is retained, which helps to keep critical speeds out of the pump running speed range.

Experience shows that the transverse hole design can be used on long or short bushings, throttle bushings, center bushings, or wear rings. Boulden, the leading provider of PFA/CF composite and developer of the PERF-Seal® design, has published case studies of boiler feed pumps upgraded with this design. Similar case studies dealt with amine charge pumps where the upgraded design again showed significantly reduced vibration.

8.3.1 Working with competent pump rebuilders

It is recommended that facilities and/or reliability professionals planning to overhaul a multi-stage pump work closely with cooperative companies. Cooperative OEMs and independent pump rebuilders show interest in systematically upgrading the reliability of the client's pumps. They provide advice instead of aiming to maintain "business as usual" attitudes.

All parties should consider available upgrade opportunities; utilizing center bushings with an engineered pattern of radial holes to equalize the pressure between the O.D. and I.D. of the bushing is among them. In some applications the upgrade will pay for itself in less than four months.

8.3.2 Results of an authoritative long-term field study made by a user

Life expectancy data for Vespel® CR-6100 can be inferred – with complete confidence – from a long-term efficiency study (Table 8.1) performed by a competent

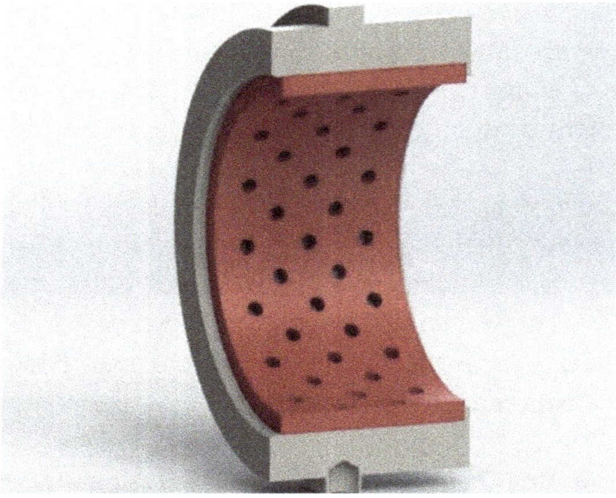

Fig. 8.3: PERF Seal®, half-segmented view (Source: Boulden Company, Conshohocken, PA, and Ellange, Luxembourg).

user. Due to the location of the user's operations, they pay higher-than-average power prices. Most of their pumps are horizontal multi-stage API Type BB3 pumps in refined hydrocarbon service. Many of the pumped fluids are flashing, light, low viscosity products. The pumps tend to be quite large, ranging from 500 kW (670 hp) to 2 MW (2,700 hp). Because of the high operating costs, the user routinely conducts tests to track pump efficiency over time.

The user site had collected efficiency data on three different material pairings for their pump wear rings:

- Cast Iron vs. 420 SS – the original pump design from the 1980s
- NiCr hard-coated rotating and stationary rings – 1990s upgrades
- Vespel® CR-6100 vs. 420 SS wear rings – current upgrades

The long-term efficiency in Table 8.1 tracks the efficiency over eight years for the different material combinations, averaging the results from dozens of pumps running each material combination. It summarizes data comparing the recorded efficiencies to the baseline efficiency of a pump using metal parts with API standard clearance.

Pumps using Vespel® CR-6100 started operation with wear ring clearance reduced by 50% compared to the pumps with metal wear parts. For horizontal multi-stage pumps, the observed efficiency improvement attributable to a 50% reduction in clearance provided an average initial efficiency gain of about 4%; this explains why the baseline efficiency of pumps with Vespel® CR-6100 wear parts is higher.

Table 8.1: Efficiency trend of Vespel® CR-6100 wear rings compared to baseline efficiency at API standard clearance.

Material combination/pairing	Baseline efficiency	Efficiency after 4 years	Efficiency after 8 years
Cast Iron vs. 420 SS	0%	−3.7%	−5.3%
NiCr hard-coated wear rings	0%	−1.4%	−2.4%
Vespel® CR-6100 vs. 420 SS	**+4%**	**+2.4%**	**+1.1%**

Leading up to 2019, eight years of collected data showed that the amount of wear of the Vespel® CR-6100 parts was in line with the hard-coated wear rings. Both Vespel® CR-6100 and hard-coated wear rings showed significantly less wear than cast iron wear rings.

8.3.3 Life cycle cost calculation

From an examination of the life cycle cost for each material pair or combination, it is evident that Vespel® CR-6100 was an excellent investment in the horizontal multi-stage pumps at this user plant. Assuming a typical industrial power price of $0.06/kwh, full-time operation, and a pump power consumption of 1 MW (1,340 hp), we find the following eight-year operating costs compared to baseline efficiency:

Cast Iron: $156,000; referring to additional operating costs over eight years
NiCr hard-coating: $60,000; referring to additional operating costs over eight years
Vespel® CR-6100: $106,000; anticipated cost savings over eight years

8.3.4 Seven rules for using DuPont™ Vespel® CR-6100

The Boulden Company has proactively condensed their many application bulletins into a seven-item list of guidelines for pumps, which they (or the user) can apply to just about any fluid machine. Following these seven rules gives user-purchasers confidence that they are using Vespel® CR-6100 properly and thereby improve pump reliability without compromise.

- Stay below the temperature limit of 260 °C (500 °F)
- Use PFA/CF only in stationary parts, mounted in compression
- Use Perf-Seal® design for throttle and center bushings of multi-stage pumps [2, 3]
- Install the Perf-Seal® component against a fixed shoulder to ensure it cannot be dislodged by differential pressure

- Avoid extremely abrasive services such as slurries, bottoms, or slops
- Verify that press fit, clearance, and axial length of part comply with the Boulden Installation Guide
- Ascertain that the pump rotor turns freely when the pump is assembled and ready for commissioning.

What we have learned
- PFA/CF materials are among the most valuable and immediately beneficial fluid machinery upgrades
- Work with competent suppliers, equipment OEMs, and fluid machinery rebuilders to implement component upgrades using PFA/CF components
- Look at payback achieved. Demonstrated rapid paybacks quickly make up for the higher initial cost of Vespel® CR-6100
- The merits of PERF Seals® (Figs. 8.2 and 8.3) have been demonstrated in many field applications on multi-stage pumps
- Power savings are significant. Long, flexible shafts benefit from the Lomakin effect generated at the wear rings. Used in conjunction with the PERF Seal® design, the resulting benefits will be even higher. They will also last beyond eight years.

References

[1] Aronen, R. and Russek, M. (2011), "Evaluating the Long-Term Performance of Composite Wear Rings," Proceedings of the 27th International Pump Users Symposium, Turbomachinery Laboratory, Texas A&M University, College Station, Texas.

[2] Aronen, R., Plaizier, D. and Sinclair, D. "Building a Better Boiler Feed Pump," Pumps and Systems, November 2007.

[3] Aronen, R. and Bothum, L. "Upgraded Boiler Feedwater Pump Improves Efficiency and adds Flexibility," POWER, September 2017.

Chapter 9
Oil rings are the weak link in lubricant application

Bearing topics and lubricant application topics overlap in fluid machines and, especially, in process pumps. Although oil-free pumps with gas bearings and magnetic bearings exist, their market share is far less than 1%. Accordingly, this text does not deal with them. The main issue limiting the reliability of the power end (bearing housing) of existing pumps is that many of them incorporate the least initial cost features. Only a relatively small number of process pumps are designed and sold with provisions ensuring that a proper amount of clean lubricant is consistently reaching the bearings [1].

Many pumps today, and pumps being offered for the foreseeable future, will benefit from thoughtful upgrading; simply repairing these pumps and restoring the mechanical assembly to the as-bought condition will not reduce failure risk.

The relevance of this statement is best understood by pondering over the disappointingly high rate of *repeat* pump failures in the industry. Repeat failures can occur only if (a) the root cause of a problem has not been found or (b) if the root cause is known but someone in charge is deliberately not taking remedial action.

Among the often overlooked fundamental causes of repeat failures of process pumps, we find cage-induced windage, which is the unidirectional blower effect of slanted ball separators (also called "cages") in angular contact bearings [1, 2]. The effect of this windage on oil flow illustrates the interdependence of bearing design and lubrication matters. Needless to say: Concerns for bearing initial cost, and the often misguided desire to standardize or consolidate product selection are responsible for bad bearing choices. Not all pump designers are aware of this fact; this part of our text brings it to our attention for an important reason: Best-in-class (BiC) pump users make it their business to upgrade the design of the power ends of pumps undergoing repeat bearing failures. BiC users move beyond mere replace-in-kind maintenance efforts; they eliminate the root causes of failures.

9.1 Lubricant level and oil application

Oil bath, also called oil sump lubrication, is one of the oldest and simplest methods of oil lubrication; only grease lubrication is older than oil bath lube. The bearing rolling elements pass or "plough" through a portion of this oil sump as the shaft rotates (Fig. 9.1). Oil bath lubrication is feasible unless (and until) too much frictional heat is generated by the oil-plowing action of rolling elements at high speeds. Because heat reduces oil film thickness and strength and greatly accelerates the rate at which oil oxidizes, the oil bath lube method is avoided on process pumps whenever DN, the inches of shaft diameter (D) multiplied by shaft revolutions-per-minute (N) exceeds 6,000.

https://doi.org/10.1515/9783110674156-009

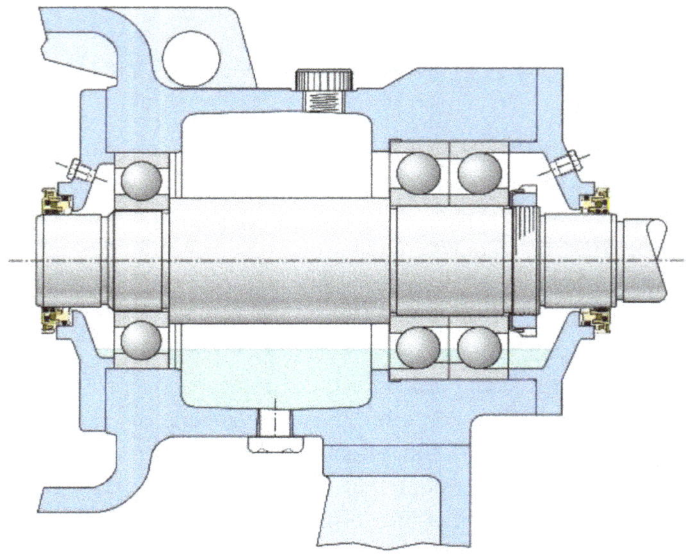

Fig. 9.1: A typical pump bearing housing with oil level reaching to center of lowermost rolling elements. In this style of bearing housing, keeping the DN-values below 6,000 reduces the risk of oil overheating. Magnetic face type bearing protector seals are shown at each end (Source: AESSEAL, Inc., Rotherham, UK, and Rockford, TN).

9.1.1 Illustrating the DN approach

It can be reasoned that a two-inch shaft at 1,800 rpm, with its DN value of 2 × 1,800 = 3,600, would operate in the suitable-for-oil-bath zone where lubricant would reach the rolling elements and where, as shown in Fig. 9.1, no oil rings would be needed. However, pumps incorporating a two-inch shaft operating at 3,600 rpm (DN = 7,200) would use oil rings (sometimes called "slinger rings") or flinger discs (Fig. 9.2) to lift or spray the oil from a sump with its level purposely maintained below the bearing. Less frictional heat results from lower oil levels (Fig. 9.2 and 9.3) than from an oil level reaching the center of the bearing elements (Fig. 9.1).

The DN limit of 6,000 is an experience-based value which assumes real-world conditions of misalignment and a host of other factors that make the actual operating pump environment different from short-duration test stand practices. Be prepared to have the limiting value disregarded and even disputed by pump manufacturers for the usual reasons. Their dispute may not always be attributable to sinister motives; it may simply be the result of being uninformed or indifferent to the fact that ten hours of test stand observation of a perfectly horizontally installed pump with a properly adjusted constant level lubricator differs considerably from the pump's performance in the field. There, in the field installation, the same pump is exposed to rain and sunshine, freezing cold and blistering heat. At the client's site, it is expected to run well with badly contaminated oil and a shaft slightly out-of-horizontal, causing the oil ring to travel downhill. As the oil ring then scrapes the inside of the bearing housing many times per hour, the oil ring is abrading. Metal flakes end up in the oil and bearings are compromised. This scenario explains why pump bearing lives rarely ever come close to the 40,000-hour L-10 rating. The L-10 life is the full-load operating time when 10% of the bearing population show indications of defects developing or progressing.

9.2 Issues with oil rings

A bearing housing with a deliberately lowered oil level and intended for DN-values in excess of 6,000 was shown in Fig. 9.2. Bearings with DN-values in excess of 6,000 will require the addition of either a flinger disc (as shown in Fig. 9.2) or an oil ring (Fig. 9.4) or similar lube application component, to dependably lift or spray-feed oil into the bearings.

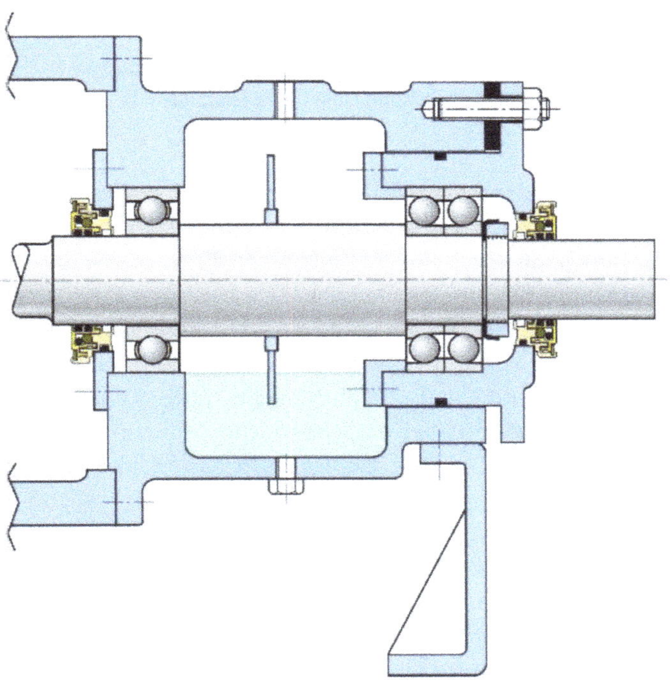

Fig. 9.2: Bearing housing with oil level lowered to accommodate high DN values. Here, a flinger disc (often called an "oil thrower") lifts or sprays oil into the bearings.

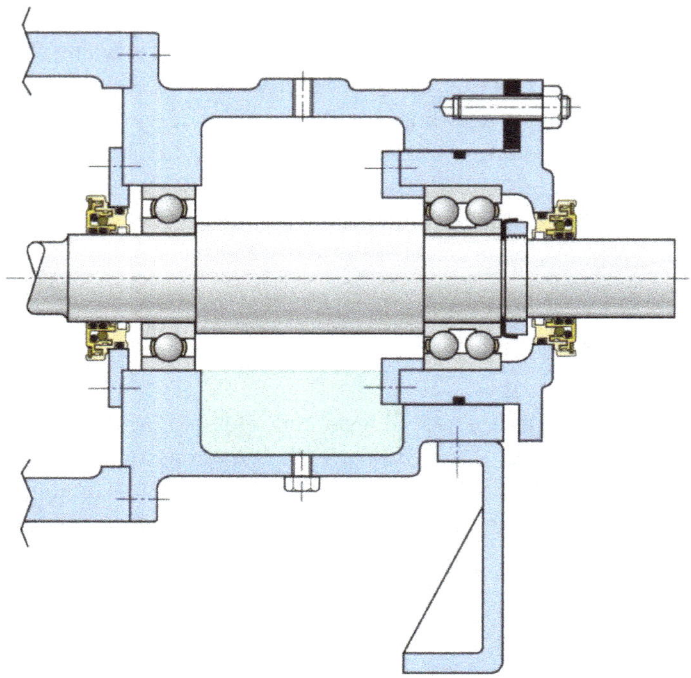

Fig. 9.3: The oil level in this bearing housing has been lowered in anticipation of DN-values in excess of 6,000. An oil ring or other oil application method must be selected next.

But oil rings are potentially vulnerable components. They will not interact the same with lubricants of different viscosities, or at different depths of oil immersion. Unless used on perfectly horizontal shaft systems, oil rings will run downhill and then often contact the bearing housing-internal parts, as easily visualized in Fig. 9.4. We should note that already in the 1970s, a then prominent U.S. pump manufacturer recognized the pitfalls of using oil rings. Its advertising literature pointed out that this company's reliable pumps incorporated an "anti-friction oil thrower ensuring positive lubrication to eliminate the problems associated with oil rings" [2]. Some European-made pumps have avoided the drawbacks of oil rings by incorporating flinger discs; they have had good success with flinger discs (Fig. 9.2) for many decades. We suspect these are not used because, due to their diameter requirements generally exceeding the bearing housing bore diameter, the bearing will have to be accommodated in a removable cartridge. Manufacturing and mounting this cartridge will add to the cost of a process pump. Regrettably, the initial cost is the predominant concern of those unable or unwilling to do even the most elementary payback calculation.

Fig. 9.4: Bearing housing with an oil ring.

As we carefully examine Figs. 9.1 through 9.3, we again note an important omission: For failure risk reduction, pressure and temperature equalizing passages – although not shown here – must be provided at each bearing. Two issues will arise without these:

(a) A small amount of oil will be trapped between bearings and the adjacent bearing housing protector seals (BPS). The oil will overheat and adversely affect bearing life (Sources: [1, 2]).

(b) The pressure in the central space between the bearing on the left and the two bearings on the right may differ from the pressures acting between this large central space and the much smaller spaces between a bearing housing protector seal and its adjacent bearing. This often causes oil leakage and oil loss past the BPS to the ambient environment.

9.2.1 Oil ring deformation

In order to resist deformation (becoming slightly oval) while operating, oil ring fabrication must include an annealing step to relieve stresses. Without this annealing, many oil rings will become progressively more oval in shape. Also, oil rings tend to become progressively more unstable as DN-values approach or exceed 8,000. Instability means that the oil rings skip, skew, misalign, and abrade. It would be a highly desirable maintenance practice to measure the new ring at assembly and to again measure it at the time of repair. The width difference represents lost metal;

the lost metal will become an oil contaminant and cause the bearings to fail prematurely.

To get oil rings to function as designed, the shaft system must be near-perfectly horizontal. Ring immersion in the lubricant must be in the right range – usually in the range of 8–11 mm (approximately 13/32 inches) below the oil level. Moreover, to avoid ring abrasion and dangerous oil contamination, ring eccentricity must be within 0.002 inches (0.05 mm), and surface finish should be reasonably close to 32 and, at most, 64 RMS. Oil viscosity should be in close range of typical ISO VG 32 properties, and temperatures must be in the moderate range [3]. These different and equally important parameters are rarely all within their respective desirable ranges in an operating plant. If several of the individual parameters are just "borderline acceptable" oil rings will intermittently malfunction. That, then, is another elusive failure cause described in [4] and implied in [5].

9.2.2 Upgrading from inexpensive oil rings

It has been pointed out that grooved oil rings perform slightly better than the plain or flat oil ring variety [5]. Also, certain plastics seem to perform better than the brass or bronze rings typically found in pumps [6]. It is more commonly known that oil rings work better with ISO Viscosity Grades 32 or 46 oil as compared to ISO VG 68 oil. (Important: A well-formulated synthetic Grade 32 may be required to suit both the constraints of oil rings and the needs of bearing housings that contain both a rolling element and a sleeve [sliding] bearing. Refs. [7] and [8] have more details on the issue.)

Here is the bottom line for the truly reliability-focused: Because oil ring behavior is very difficult to control, some reliability-focused purchasers try to avoid them. These users often specify and select pumps with large diameter flinger discs (Fig. 9.2) in DN-applications where lower oil levels are needed. *Small* diameter flinger discs are sometimes used in oil bath applications. Labeled "oil throwers" in Fig. 9.5, small diameter discs are intended to simply prevent temperature stratification of the oil. Without these discs, hot oil would tend to float at the top of the oil sump. However, due to their relatively small diameter, they are not the equivalent of large diameter flinger discs. Whereas the latter throw oil into the bearings for the purpose of lubrication, the former are used in bearing housings with oil levels reaching the center of the lowermost rolling elements. Small diameter discs merely serve to keep oil temperatures uniform in pumps with DN-values below 6,000.

9.3 Pressure and temperature balance in bearing housings

A few paragraphs earlier, we made the point that deficiencies in pressure and temperature balance are among the elusive failure causes in pump bearings. The issue is worth repeating here. Note that the double-row radial bearing in Fig. 9.5 was – in the 1960s – called a "line bearing." Take a careful look also at the balance holes in the bearing housing near the top of each of the two bearings. The then prominent manufacturer Worthington Pump Company knew the importance of achieving pressure balance throughout the entire pump bearing housing. Oil leakage risk past the oil seals (they were lip seals in this very old design) was greatly reduced or fully avoided by keeping all pressures equal. Pressure balance is important also because it largely neutralizes the potential windage effects of certain slanted (angular contact) bearing cage orientations.

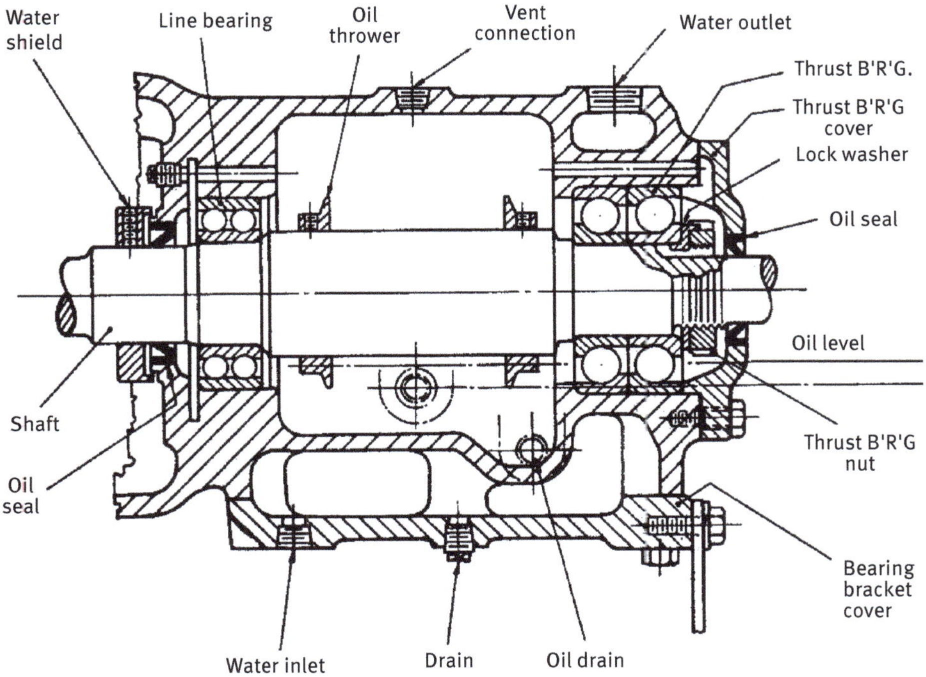

Fig. 9.5: This 1960-vintage bearing housing uses "oil throwers" for keeping the oil at uniform temperature. Note pressure balance holes provided at the top of radial and thrust bearings (Source: Worthington Pump Installation and Operating Manual, ~1966; also [1]).

Flexible flinger discs were briefly tried to enable insertion in the simplest bearing housings. In these trials, the bearing housing bore diameter was slightly smaller than the flinger disc diameter. It was determined that, unless the elastomeric (flexible) material is carefully chosen for a combination of flexibility, strength, and chemical resistance, the user would be better off specifying the preferred *solid* steel flinger discs. In that case, the more accessible of the two pump bearings may have to be cartridge-mounted. With the cartridge design depicted in Fig. 9.2, the effective bearing housing bore (i.e., the cartridge diameter) must be larger than the steel flinger disc and therefore allow insertion of a steel flinger disc with greatest ease.

Of course, providing such a cartridge will add to the cost of a pump. Yet, in virtually every conceivable instance, the incremental cost of the cartridge design will be low compared to what it costs to repair a pump. The incremental cost of a cartridge is always less than the cost of even a single repair. The same payback argument can be made for balance holes or passageways; they are needed to ensure housing-internal pressure and temperature equalization and add pennies to the cost of a pump. Internal pressure and temperature balance between the central volume of the bearing housing and the spaces between bearings and housing end caps are essential to proper oil delivery and leakage prevention from the bearing housing end covers. However, in the years from 1990 to 2020, this requirement appears to have been disregarded in some pump models. Still, a well-written specification insists on balance holes.

In some instances, it would be best to provide pressure equalization passages at both top and bottom of the bearing bore. An oil return passage would be positioned at or near the 6 O'clock location of a bearing periphery to prevent oil from getting trapped behind a bearing. Trapped oil will overheat and turn into high-viscosity tar or some other undesirable form of carbon-containing residue. The lack of these passages is one of several explanations for oil leakage, overheated oil, and premature bearing failure.

Overheated oil and/or oil contaminated with slivers of black O-ring material will result in "black oil." This is a hugely important comment since old-style bearing protector seals were often designed with a dynamic O-ring moving up and down between a sharp-edged O-ring groove in the stator bore and an opposing sharp-edged O-ring groove in the rotor periphery. Some purchasers were wise to the risk; they reported removing this "dynamic" O-ring before installing this widely used "bearing isolator" in their rotating equipment.

Note also that the 1960-vintage bearing housing of Fig. 9.5 shows lip seals and water-cooling provisions. The lip seals shown here would no longer be acceptable in API-style pumps; the modern bearing protector seals discussed earlier in Chapter 4 and shown here in Figs. 9.1 through 9.3 would be used instead. Although cooling water passages are pointed out in Fig. 9.5, cooling water was deleted from pumps with rolling element bearings in the early 1970s [4, 8]. Cooling water is not required on pumps equipped with rolling element bearings. If a pump manufacturer advocates the use of cooling water on such bearings, it would be wise to question the design. Chances are that bearings have an excessive interference fit or the entire bearing housing is simply too small. A suitably dimensioned oil sump is needed for rotating equipment and, in the case of pumps, making up for an inadequate design by shifting the cooling responsibility to the user's cooling water system is not tolerated by BiC user companies. API 617 alludes to this fact, and we will say a bit more about the issue with cooling water later in this chapter.

9.4 Things to watch with constant level lubricators

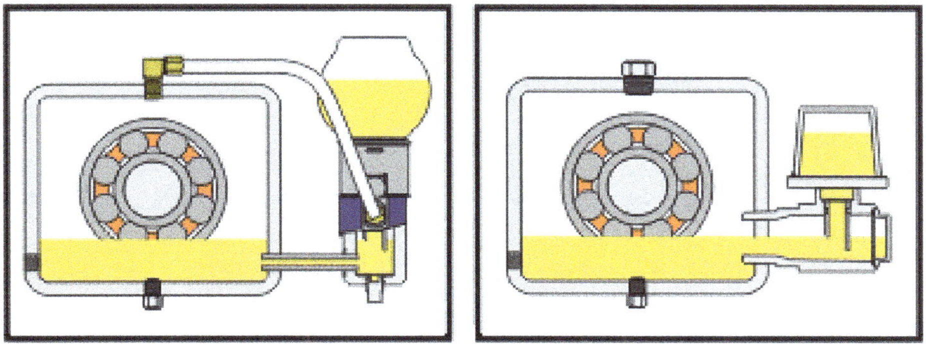

Fig. 9.6: Popular Constant level lubricators, closed system "Optomatic" on left, "Watchdog" on right (Source: TRICO Corporation, Pewaukee, WI).

Whenever one opts for the implementation of either oil mist or oil sprayed directly into the cage of a rolling element bearing, one will have reached a true milestone: One can now remove and permanently discard oil rings and constant level lubricators (Fig. 9.6). Bearing housing maintenance and failure risk are greatly diminished once lubricant application no longer depends on these two attention-requiring oil delivery components. This is why oil mist systems have been so successful at BiC companies.

Nevertheless, it can be readily acknowledged that both can work well if the owner-purchaser strictly follows the procurement and installation specifications for oil rings and constant level lubricators. For oil rings, good procurement specifications include stress-relief annealing before finish machining, dimensioning thoughtfully guided by the intended lubricant viscosity, proper depth of immersion in the oil, and ascertaining shaft horizontality. Constant level lubricators should be the pressure-balanced type, shown in Fig. 9.6 (left side). Like other constant level lubricators with glass bulbs, they require time-based replacement of the oiler bowl assembly due to age and temperature-related degradation of the caulking material.

9.4.1 Where to place a constant level lubricator

There is never a substitute for following the laws of physics, hydraulics, and plain common sense whenever we deal with machinery. The value of reasoning on matters and understanding how components and subassemblies work should never be underestimated, as the example will show. Consider it just one of the hundreds that highlight why Best Practices should be consistently implemented [9] and why correctness is best verified by conducting periodic field audits.

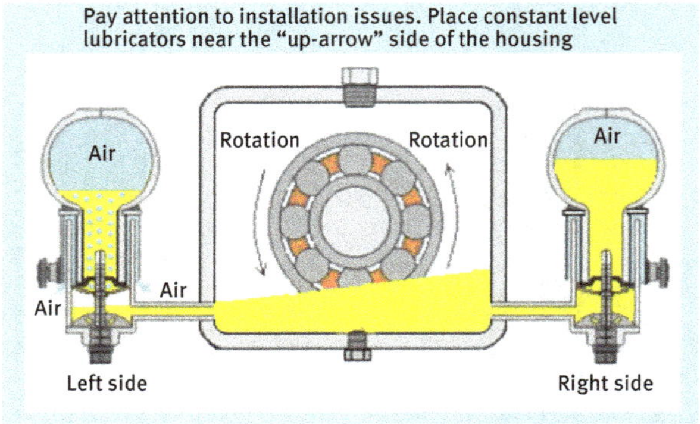

Fig. 9.7: Constant level lubricators should be mounted on the rotating shaft's "up-arrow" side. Mounting on the wrong side tends to allow air to get aspirated [9].

Fig. 9.7 illustrates why constant level oilers should be mounted on the rotating shaft's "up-arrow" side [9]. Mounting on the wrong side (the "down-arrow" side) tends to allow aspiration of air. As we further reason on the matter, we can show that plants that have the constant level lubricator working well on the *wrong* side will have moved the wing nut supporting the glass bulb to a height setting that was *not intended* by the designer. In other words, for a deviation from a proper instruction to still work well, there must be a compensating deviation elsewhere. To use a fitting analogy: If, in a country where one drives automobiles on the right side of the road, we decide to drive on the left, oncoming traffic will have to compensate by (temporarily) driving on their left whenever they see us approaching. While in that one instance two wrongs will admittedly make a right, we would be reluctant to defend the risk-prone practice.

9.4.2 Two lubricators?

Another refinery is apparently unaware of the "up-arrow" rule. Between-bearing pumps use constant level lubricators at both bearing housings, but each should be positioned on the same side, the "up-arrow" side. We will have to know the direction of shaft rotation to identify which one of the two lubricators in Fig. 9.8 should be relocated to the other side of the pump bearing housing.

It should be noted that all three references (below) explain the "up-arrow" rule and other zero-cost reliability-enhancing steps. A few leading plants have purchased relevant technical texts for their maintenance-technical work force members. The maintenance technicians are then asked to read a single page each day and highlight a "Comparing Practices Ledger" where local practices differ from those described as BiC facilities in the cited references. The benefit-to-cost ratios resulting from this page-per-day reading and comparing one's own practices to those of BiCs are intuitively evident.

Fig. 9.8: One of the two lubricators is mounted on the wrong side of its bearing housing. We will have to know the direction of shaft rotation to identify which one to relocate [3].

They have consistently exceeded all expectations. Whenever reliability professionals have to stay at home for unspecified reasons, smart managers often ask them to stay productive by reading one or more of the many reference book listed throughout this text. The home-bound technical employees would be instructed to use their word processors and jot down "us" versus "them." They would start typing, on sheets of paper or in a computer file, two contrasting practices in two-column format. In one column would be "how we do it here," in the other column they would list the practices and work procedures at BiCs. The contrast is often revealing.

9.5 Cooling water is never needed on pumps with rolling element bearings

Although cooling is still *found* in pumps with high speed and heavily loaded rolling element bearings, cooling of the oil is rarely, if ever, *needed*. In fact, cooling a rolling element bearing can represent more than the mere waste of a precious resource – water. Not only is water cooling of no benefit in installations with properly configured bearing housings, but cooling the oil can be downright detrimental to bearing life. Here's why:

Suppose a cooling jacket restricts the bearing outer ring from free thermal growth in all radial directions while the bearing inner ring heats up and grows, causing bearing internal clearances to vanish. An excessive preload could result (ref. [1] deals with bearing preloads in more detail).

Similarly, immersing cooling coils in the oil will cool not only the oil but also the air in the bearing housing. Such cooling could promote moisture condensation from the air and cause harmful oil contamination. Therefore, cooling water has been deleted from every pump with rolling element bearings at Best-in-Class (BiC) locations [1–3]. API 610 asks that pumps with rolling element bearings use cooling water only if the pump model requires it. To the author, such requirements would rouse suspicions that the pump model at issue has been designed with insufficient oil capacity in the bearing housing. Designs with too small an oil volume should certainly trigger a close review and possible disqualification of such a pump.

Since the late 1970s there no longer is cooling water on bearing housings of pumps with operating (fluid) temperatures up to and including 752 degrees F (400 degrees C) in modern BiC oil refineries. Sadly, some users are stuck in doing things the way they've always done things: They wait for the next failure and then react to it, over and over again.

Because cooling water ports are shown on a pump drawing, the user is led to believe that such cooling is needed and helpful. Commenting again in Fig. 9.5: When it was discovered that cooling is no longer needed, BiC companies began to leave these cooling water drains open. With modern synthetic lubricants and properly selected rolling element bearings, cooling is simply no longer required in well-designed process pumps. We want to restate this fact and impress on the reader: Whenever high-performance synthetic lubricants of the correct viscosity are used, there is no longer a need for cooling water on pumps equipped with rolling element bearings. These findings have been validated thousands of times and are not dependent on lube application method.

However, with *sleeve bearings*, cooling is still used in order to maintain optimum oil viscosity through close temperature control. This cooling provides a reasonably stable environment for oil rings in pumps equipped with sliding (*sleeve*) bearings. Circulating systems ("pump-around systems") are the primary choice for large pumps that incorporate these bearings.

Very large process pumps (~400 kW and larger) use sleeve bearings and circulating oil systems. In circulating systems, the oil can be passed through a heat exchanger before being returned to the bearing. Pressurization is needed to move oil through filters and exchangers, but the bearing itself is rarely fully pressurized. In some sleeve bearing systems, oil rings are used to lift the oil and deposit it on the shaft surface. In some other sleeve bearing systems, an oil spray from suitably placed nozzles is directed to oil grooves with good effect and very high reliability [1, 2, 7, 8].

9.5.1 Redesigned bearing environment

As to major process pumps with rolling element bearings, there now exist clear recommendations to redesign bearing housings. Best-available technology housings would include a shaft-driven pump (Fig. 9.9) in the redesigned housing space or layout. Oil from the sump would be pressurized and sprayed into the rolling element bearings.

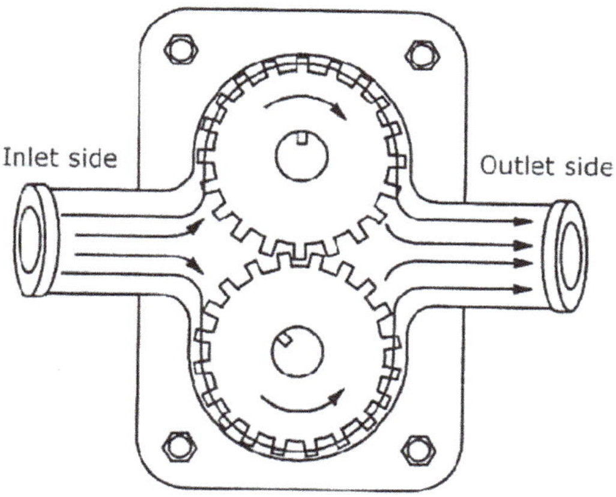

Fig. 9.9: Open source illustration of a gear pump placed in the bearing housing oil sump and driven by skewed gears (not shown).

All major bearing manufacturers consider an oil spray to be the most advantageous lube application available. Upgrading to self-contained oil spray application would eliminate troublesome oil rings and maintenance-intensive constant level lubricators. Such upgrades would be of great interest to the truly reliability-focused.

As early as six or seven decades ago, a group of reliability engineers focused on lubrication technology and studied the merits of synthetic lubricants. Without these synthesized hydrocarbons, aerospace endeavors would have moved at a snail's pace. About 50 years ago, reliability professionals began familiarizing themselves with different lubricant application methods found in process pumps. These professionals learned that all manufacturers of rolling element bearings give ranking numbers to different lubricant applications.

In 1980 and today, in 2019, all major bearing manufacturers agree that a jet of oil sprayed into bearings is the most effective way of lubricating rolling element bearings. Oil mist lubrication, comprised of a 200,000:1 air-to-oil mixture (by volume), is usually highly ranked as well.

As mentioned, pressured lubricant ("oil jet" or "oil spray") probably represents optimized lubrication to process pumps equipped with rolling element bearings. Favorable comments on external pump-around lubrication (Figs. 9.10 and 9.11) notwithstanding, the use of separately powered external oil pumps to pressurize and then spray lube oil into bearings may not be economical. However, developing and then positioning a small oil pump inside a process pump's bearing housing is feasible and is anticipated to yield rapid payback. Located inside or mounted directly on

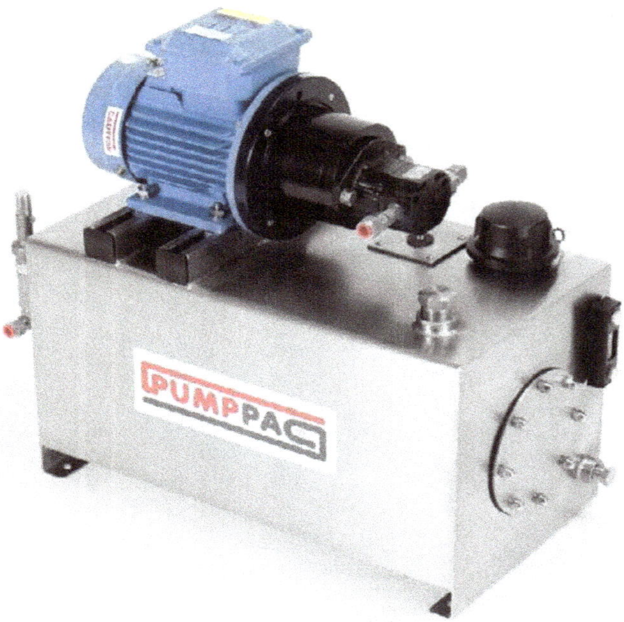

Fig. 9.10: Closed-circuit pump-around unit for circulating oil to bearings, or barrier fluid to dual mechanical seals (Source: AESSEAL, Inc., Rotherham, UK, and Rockford, Tennessee, USA).

the bearing housing, it could be driven from the main pump shaft via skewed bevel gears. The small positive displacement spur gear pump (Fig. 9.9) or a functionally similar motor vehicle crankcase oil pump would take its flooded suction from the oil in the sump and would force the oil from the small pump's discharge port through spray nozzles into the pump bearings [3].

9.6 Pump-around units: the higher-order upgrades

Two examples from the author's consulting experience add to our conversation on the drawbacks of oil rings and inadequately sized bearing housing dimensions. First, the effects of out-of-roundness of these rings. Oil rings are risky, at best.

9.6.1 Exceeding recommended out-of-roundness tolerance is risky

An aluminum smelter in the United States reported frequent bearing distress in their boiler feed water pumps. Because more than one pump had to run to satisfy demand and pumps suffered from random failures, the author measured the parts from a dismantled pump. He found the oil ring 0.061 inches out-of-round. This exceeds the allowable tolerance given in ref. [5] by a factor of 30. An oil ring with such inner diameter out-of-concentricity will probably act like a speed-up and slow-down pendulum. At the aluminum smelter, operating four pumps in parallel no doubt contributed to bearing distress at the facility's boiler house. The performance curves or these pumps were flat in the region close to pump shutoff (zero flow), at which time the thrust bearings were highly loaded, or perhaps even overloaded. Inadequate oil flow meant less heat was carried away at that time, and encountering bearing distress was just a matter of time.

Our second example relates to an oil refinery in the West Central United States. Within two weeks of attempting to start up a new process unit, one of the refinery's newly designed hot service process pumps experienced several bearing failures. Matters got so bad that an observer had to be posted at the pumps to monitor bearing housing temperatures. With no time to waste on an in-depth root cause failure analysis, the author recommended immediate procurement of one of the two plausible retrofits shown in Figs. 9.10 and 9.11. The pump-around unit in Fig. 9.10 is designed and built for duty in dual seal systems where barrier fluid must be pumped around, pressurized, and, in some cases, cooled.

Fig. 9.11 represents the second option, a self-contained oil mist unit of the type and style used whenever plant-wide systems were not available because their cost was not included in the original budget. Using either one would allow scrapping the oil rings and constant level lubricators. The customary breather vent located at the top of a bearing housing would also be deleted.

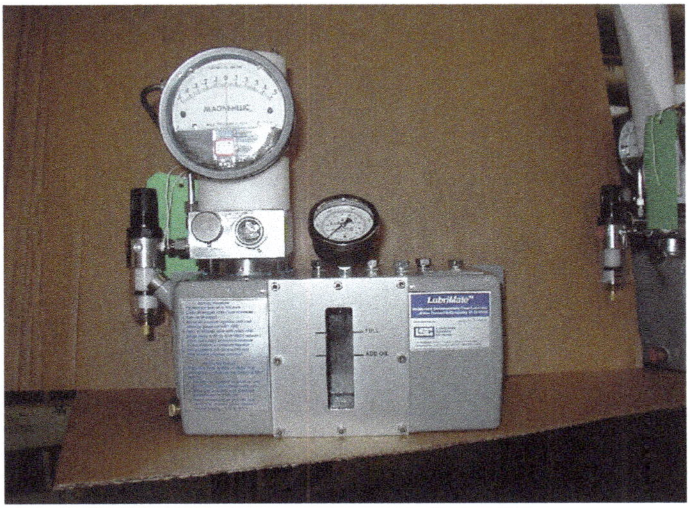

Fig. 9.11: Self-contained oil mist supply unit for equipment where lube supply and application need to be upgraded (Source: Lubrication Systems, Inc., Houston, Texas).

Within an hour of this diagnosis, we located a leading oil mist provider company which, days earlier, had taken a small oil mist unit (Fig. 9.11) in trade for a much larger unit. The oil mist provider's service truck delivered the unit to the purchaser's location and, two days later, the pre-owned unit was hooked up to the troublesome pump. Although new, the process pump almost certainly had been mis-designed with too small an oil sump. It took only about six hours to connect the oil mist unit; all probable sources of distress on this process pump – defective oil rings, misadjusted constant level lubricators, and/or inadequate oil flowrates – were thereby eliminated.

Of course, there are ways of leaving behind many of the concerns just described in Chapter 9. One such method of avoiding issues is to consider modern canned motor pumps; these are the subject of our next chapter.

What we have learned

- Late nineteenth-century lube oil delivery methods are not always harmonizing with early twenty-first-century reliability goals
- Oil rings are working well if they have been stress-relieved and manufactured not to exceed 0.002 inches out-of-roundness
- Pressure and temperature equalization are often lacking from pump bearing housings. Proactive Reliability Professionals insist on such equalization
- Oil rings will malfunction unless installed on fully horizontal shaft systems. If allowed to run downhill and to then contact housing-interior surfaces, they are prone to abrade
- For correct functioning, proper oil ring immersion is important, as is maintaining correct lubricant viscosity
- Flinger discs are an upgrade; pump-around units and small oil mist units qualify as higher-order upgrades
- Constant level lubricators are (a) direction-sensitive and (b) require periodic replacement
- While at first glance it seems reassuring that 90% or even 95% of pumps with oil rings seem to operate well, 5% of 3,000 pumps would equal 150 pumps with malfunctioning oil application and randomly failing bearings. That would be unacceptable to a reliability-focused organization. They would strongly advocate permanent remedies and would insist on reducing failure risks.

References

[1] Bloch, Heinz P. "Pump Wisdom," (2011), John Wiley & Sons, Hoboken, NJ.

[2] Bloch, Heinz P. and Alan Budris. "Pump User's Handbook: Life Extension," (2014), Fourth Edition, Fairmont Publishing Company, Lilburn, Georgia, ISBN 0-88173-720-8.

[3] Bloch, Heinz P. "Petrochemical Machinery Insights," (2016), Elsevier Publishing, Oxford, UK, and Cambridge, MA, ISBN 978-0-12-809272-9.

[4] Bloch, Heinz P. "Machinery Reliability Improvement," (1998), Third Edition, Gulf Publishing Company, Houston, Texas, ISBN 0-88415-661-3.

[5] Wilcock, Donald F. and E. R. Booser. "Bearing Design and Application," (1957), McGraw-Hill Company, New York, NY.

[6] Bradshaw, Simon. "Investigations into the Contamination of Lubricating Oils in Rolling Element Pump Bearing Assemblies," (2000), Proceedings of the 17th International Pump User's Symposium, Texas A& M University, Houston, TX.
[7] SKF USA, Inc. "Bearings in Centrifugal Pumps, (1995), 2nd Edition, Publication I.D. 100-955.
[8] Bloch, Heinz P. and Kenneth E. Bannister. "Practical Lubrication for Industrial Facilities," (2017), 3rd Edition, The Fairmont Press, Lilburn, GA, 30047, ISBN 088173-761-5.
[9] TRICO Manufacturing Corporation, Pewaukee, WI. Marketing Bulletin, www.tricocorp.com.

Chapter 10
Consider *today's* canned motor pumps

Before we delve into the important topic of *today's* canned motor (seal-less) pumps, here are the reasons why we opted not to include a full chapter on conventional centrifugal pumps in this book. Check index for essential information.

Pump failure statistics clearly show that mechanical seals and bearings are the two components that account for at least 80% of failures occurring in conventional process pumps. Five of the previous chapters explained how to avoid most of these failures through optimized selection of bearings (Chapter 3), better bearing housing protector seals (Chapter 4), superior selection strategies for mechanical seals (Chapter 6), installing high-value wear parts (Chapter 8), and the purposeful elimination of oil rings in pump bearing housings (Chapter 9). All of these updates are thoroughly explained and expanded in the slightly over 1,000 pages of [1] through [3]. Management summaries of these topics can be found in the 724 pages of [4].

10.1 Do we really know these pumps?

To begin our examination of canned motor pumps, we must concede an interesting fact: We often make the mistake of letting our equipment selection decisions be governed by old data, anecdotal recollections, and traditions. We illustrate the point by recalling what potential users may have read or heard about canned motor pumps in the past [5]. We might then compare the understanding we had then, with the facts that present themselves today. Today, in 2020, we know of 750 kW canned motor pumps. Today, there exist some elevated-pressure 14-stage horizontal canned motor pumps in highly successful operation.

Figure 10.1 represents a typical single-stage canned motor pump; it is thought that its highly competent manufacturer markets these as an off-the shelf design. Prefabricated modular designs allow shipping times for the chemical industry of 16 weeks; even with limited options to select the offered ranges up to 50 kW will cover most applications.

A cross-section view of the pump in Fig. 10.1 is given in Fig. 10.2; note that it incorporates a mirror-image impeller on the "free" shaft end, changing the internal flow path and adding pressure to the existing differential pressure. This forces the liquid to the discharge (high-pressure side) of the pump, in lieu of the normal flow path back to the suction side. In fact, the back impeller in Fig. 10.2 primarily compensates for the increase in temperature and vapor pressure in the motor; it thereby avoids vaporization (also known as flashing). No external piping is necessary; the full heat balance and flow happen inside the pump and do not rely on back pressure

https://doi.org/10.1515/9783110674156-010

Fig. 10.1: Typical single-stage canned motor pump (Source: Hermetic Pumpen, Gundelfingen, Germany).

Fig. 10.2: This single-stage canned motor pump incorporates an additional impeller on the free end (or "back") of the motor for services with volatile liquids (Source: Hermetic Pumpen, Gundelfingen, Germany).

or other operating influences. The thermodynamics of the pumped fluid are known to the pump manufacturer and play a role in pump design.

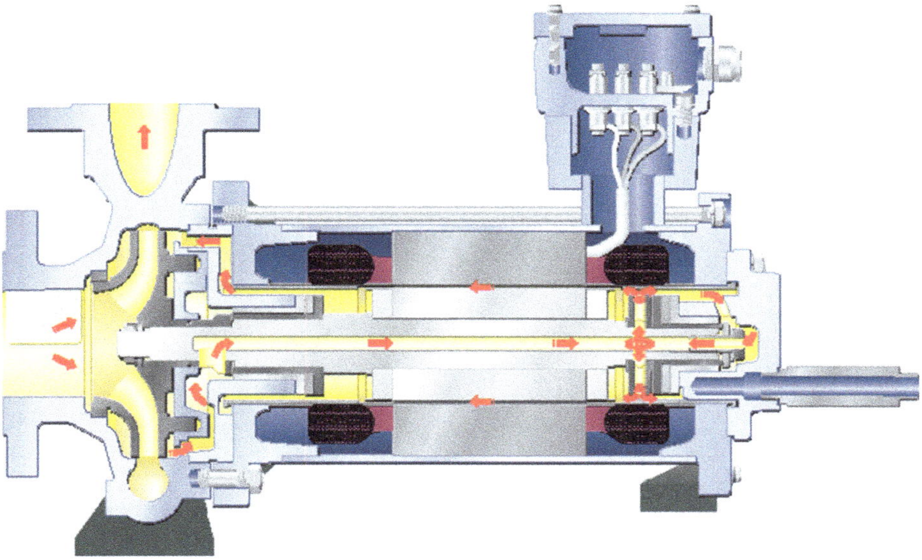

Fig. 10.3: Cross-sectional view of a single-stage canned motor pump with the internal impeller. This pump is intended for liquids with vapor pressures over 0.5 bar (7 psia) (Source: Hermetic Pumpen, Gundelfingen, Germany).

Extreme design flexibility has been achieved by at least one highly qualified manufacturer. The canned motor pumps available in 2020 cover a virtually limitless number of configurations. Again, noting the extreme variety of proven styles and models, we see a canned motor pump in Fig. 10.3 intended for liquids with vapor pressures greater than 0.5 bar or 7 psia. The canned motor pump In Fig. 10.4 is provided with a heat exchanger (cooler). This model can also handle slurries; it complies with API 685, the quality standard of the American Petroleum Institute. A 6-stage canned motor pump is depicted in Fig. 10.5.

The progress made since 1994 is staggering, and advances are obvious in just about everything, including the 8-stage "tandem" canned motor pump of Fig. 10.6.

It is accurate to state that whatever can be driven by an available canned motor can also be pumped. As mentioned earlier, even a 14-stage tandem model is presently being offered by the leading manufacturer of canned motor pumps. All models can be built in either horizontal or vertical orientation (required for low-viscosity liquids).

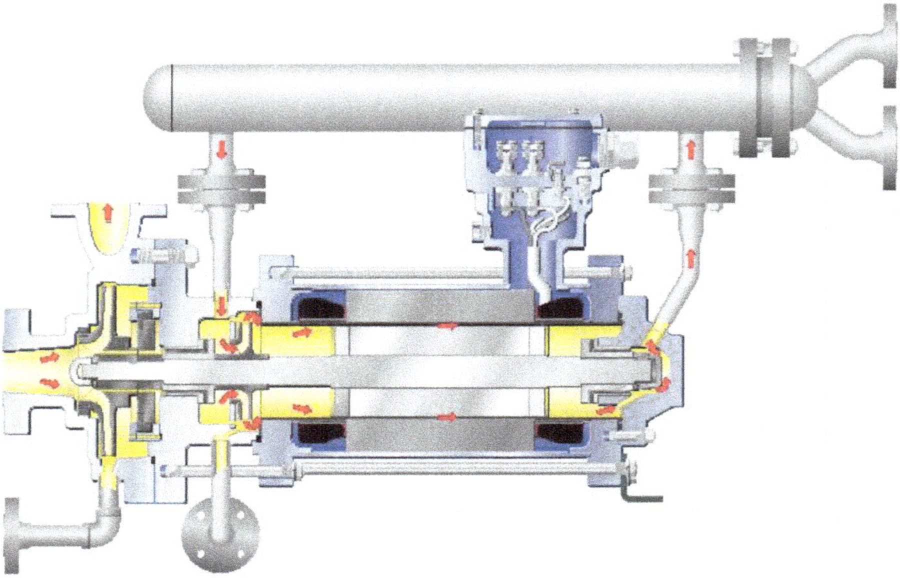

Fig. 10.4: Canned motor pump with heat exchanger (cooler); this model can also handle slurries (Source: Hermetic Pumpen, Gundelfingen, Germany).

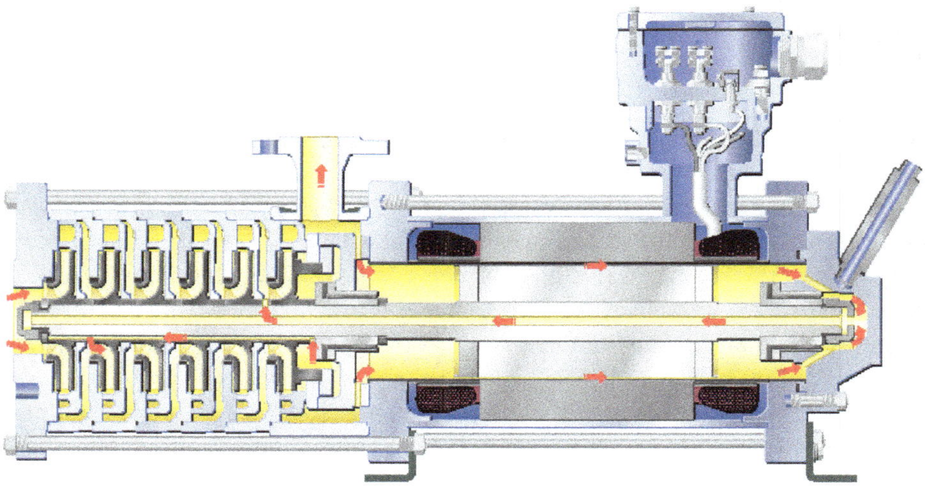

Fig. 10.5: A 6-stage in-line stage, "overhung impeller" canned motor pump (Source: Hermetic Pumpen, Gundelfingen, Germany).

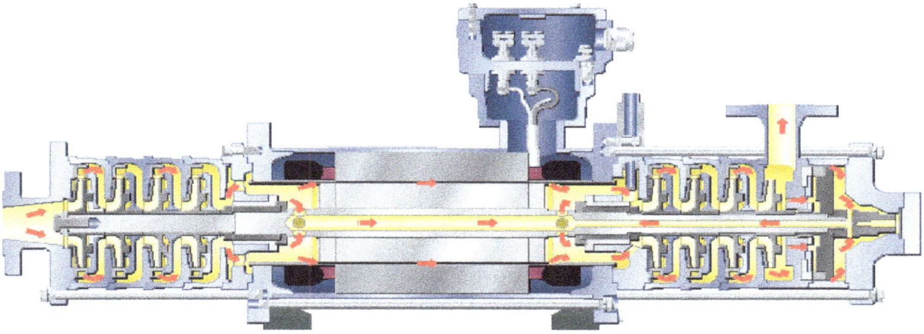

Fig. 10.6: A 8-stage ("tandem style") canned motor pump (Source: Hermetic Pumpen, Gundelfingen, Germany).

10.1.1 Designs that work

If there is liquid at the suction, any specific canned motor pump design will work with the internal flow; it does not depend on the numeric value of its vapor pressure. Still, this design applies to vapor pressures in excess of 55.16 kPa (8 psia) and reaches up to well over 1,000 kPa (hundreds of psi). In the final analysis, it is all about mechanical design pressure, not about heat or vapor pressure. It is always possible to modify the internal flow volume and pressure to compensate for any increase in the original suction conditions of liquid pressure, temperature, and vapor pressure.

To date, at least one experienced manufacturer has built and can offer canned motor pumps that pump fluids with temperatures as low as –129 °C (–200 °F), others with temperatures as high as 427 °C (800 °F). Driver power outputs have ranged as high as 750 kW. To date, the highest design pressure is 124,000 kPa (~18,000 psi). Higher pressure designs are entirely feasible although, as of early 2018, no inquiries were known to have been made exceeding this pressure rating. Temperatures as low as –200 °C were inquired and quoted as of 2019.

10.1.2 Why we should know

Becoming familiar with today's canned motor pump technology and the range of possibilities associated with these proven fluid machines is one of the many roles of reliability professionals. The fact that many profitable European oil refineries have capitalized on this technology for years deserves to be studied more closely. Not to be overlooked are "wire-to-wire" efficiencies and availabilities (i.e., uninterrupted run lengths) that for years have often matched – and sometimes exceeded – those of traditional process pump-and-driver sets. Today's state-of-technology should be explored by professionals who, in turn, would wish to inform project executives in a timely manner [5].

10.1.3 Not just one wet end manufacturer

Both the prominent canned motor designer-manufacturer and several other well-known pump manufacturing companies offer hydraulic or "wet" ends to fit the canned motors designed and manufactured by the leading provider. This flexibility with hydraulics, for example, may be interesting as a replacement for high-speed pumps driven by integral speed-up gears. If a specific high-speed pump with its associated gearing incurs high-maintenance costs, a well-designed multistage or tandem canned motor pump is often a good fit and worthy replacement.

There is undisputed evidence that canned motor pumps will often save maintenance, reduce noise, and lower the NPSH requirements. The overall efficiencies of many are matching or exceeding those of conventional pumps. Remember that the efficiencies of conventional pumps are often reported without accounting for the power absorbed by the pump's mechanical seal and shaft coupling.

What we have learned

- Canned motor pump technology has made great inroads since the 1970s. Some European refineries operate with hundreds of such pumps in one single location
- As of 2013 the various experience-backed performance parameters have covered surprisingly wide ranges; the available ranges deserve to be studied
- Significant changes in the reliability and safety-thinking at US plants have also happened since 2013 and, as of 2020, some specific US refineries report the installation of over 50 canned motor pumps at one singe plant site alone
- As of 2019, API 685 updates and revisions are in the final approval stages; the next edition of the API Standard will highlight and incorporate many experience-backed requirements and findings documented in this chapter.

References

[1] Bloch, Heinz P. and Allen R. Budris. "Pump User's Handbook: Life Extension," 4th Edition (2014), The Fairmont Press, Inc., Lilburn, GA 30047, ISBN 0-88173-720-8.

[2] Bloch/Ehlert/Geitner. "Optimized Equipment Lubrication, Oil Mist Technology, and Storage Preservation," (2020), Reliabilityweb, Ft. Myers, FL,ISBN 978-1-941872-98-7.

[3] Bloch, Heinz P. "Pump Wisdom: Problem Solving for Operators and Specialists," (2011), John Wiley & Sons, Hoboken, NJ, ISBN 978-1-118-04123-9.

[4] Bloch, Heinz P. "Petrochemical Machinery Insights," (2016), Elsevier Publishing, Oxford, UK, and Cambridge, MA, ISBN 978-0-12-809272-9.

[5] Neumaier, Robert. "Hermetische Pumpen," (1994), Verlag und Bildarchiv Faragallah, Sulzbach, Germany, ISBN 3-32968205-2.

Chapter 11
More than one problem may be at work in machines

In an earlier overview segment (see Chapter 2), we mentioned that successful failure analysis and avoidance strategies will have structure and they will be repeatable. Within the scope of this text we again remind the reader that simplified routines have been in use since the 1960s and that these techniques are anchored in two important observations:

1. All failure events can be found the seven cause categories 1(a) through 1(f); we again confirm there is no relevant eighth cause category:
   a. Design Defects (would be replicated in all identical machines or parts)
   b. Material Defects (detected by closely examining fracture surfaces, etc.)
   c. Processing and Manufacturing Deficiencies (e.g., using flawed heat treatment)
   d. Assembly and Installation Defects (inserting skewed gaskets, allowing careless piping procedures, etc.)
   e. Off-design or Unintended Service Conditions (all designs must fit into an intended operating range)
   f. Maintenance Deficiencies (neglect, using incorrect procedures)
   g. Improper Operation (disregarding the fact that nothing is indestructible)
2. Mechanical components can only fail due to "FRETT" – Force, Reactive Environment, Time, or Temperature.

As one tries to either find cause categories or, in the case of failed mechanical parts, the prevailing basic agent of the four possible parts failure mechanisms "FRETT," one proceeds with a process of elimination. The one or at most two not-eliminated cause categories or likely parts failure mechanisms will then be interrogated more closely [1].

11.1 When deviations combine

Whenever we investigate why a machine fails repeatedly and seemingly at random, it's important to remember that more than one deviation (or failure cause) may be involved. Letting deviations combine is quite risky, and if acceptance of deviations becomes the "new normal," failure risks often escalate to the point of danger.

In the great majority of machines operating in modern process plants, three principles pertain and should be kept in mind:

– When deviations combine, serious failure events are rapidly approaching
– All failures have causes. Unless causes are found and addressed, more failures will occur

https://doi.org/10.1515/9783110674156-011

- Most root causes of machine component failure can be uncovered and remedied by a properly led and trained workforce [1]

To illustrate the point, this brief chapter segment first describes a few case histories, with the first one involving a small process machine. It documents how several deviations proved troublesome in a 200 hp, vertically oriented, integrally geared high-speed low-flow (HSLF) single-stage centrifugal compressor. The machine had experienced at least five costly bearing failures in the span of two years before an experienced reliability professional was asked to participate in a structured investigation. As is so often the case, more than one deviation was found responsible for the problems encountered with this fluid machine.

11.2 Case History No. 1 and some of the seven categories of deviation explained

To get started, the reliability professional whom we will here call Troubleshooter/ Failure Analyst ("TFA") examined parts, pieces, and data after another failure occurred on a single-stage HSFL compressor. Considerable data had been collected by the equipment owner-operator and the timeline for the different failures was not easy to reconstruct.

There was much evidence that this HSLF had suffered from different issues. Moreover, it was soon evident that each failure incident had been treated along the lines of "the part that failed must be at fault." True, if defective parts fabrication can be proven, simple part replacement may be a suitable action to take. Using the seven-root-cause process of elimination, the TFA reasoned as follows:

11.2.1 Not a faulty design – (1.a)

If there is clear evidence of a faulty design, all such parts or machines would be affected. In that case, re-engineering might be appropriate. But if one redesigns "Component A", the functionality of other components may be compromised. Caution would then be the wisest course of action and a rather comprehensive reassessment of many interacting factors would be needed. Because many other machines of this type and model, were in successful service it could be concluded that the cause category listed above as 1(a) could be crossed off.

11.2.2 Quality control issues – when and where?

Remember our lead-in paragraphs which made the point that all machinery failures can be placed in just seven cause categories. Early in the chain of separate failure events the equipment owners had decided that vendor quality control (QC) during fabrication was at fault, but where was the evidence? What measurement were taken to prove which components defective? The equipment owner had *suspected* QC (quality control) events in three of the seven cause categories. But no *when, why, and where* was explained.

11.2.3 Off-design or unintended service conditions – (1.e)

Without data, all possibilities are mere hypotheses and opinions. However, the TFA insisted on *seeing data*. When he examined (and re-read) data collected by the plant's process computer, he quickly found that the small compressor had often been operated outside its allowable design flowrate or range. Time was taken to explain the vulnerabilities of low flow operation. Operator or automatic control-related deviations from the original design pointed to a relevant cause category when at least one subsequent low flow event was found in the older collection of operating data.

Low flow operation (operation in surge) means that the gas flowrate is so low that the gas volume pulsates back-and-forth as it travels from the compressor's inlet (at suction pressure) to its outlet (at discharge pressure). Pulsating flow would also explain failures of the thrust bearing, although thrust bearing distress could be the result of lube oil foaming and/or lube supply quantity and temperature having been out-of-range. A thorough review of lube oil properties was among the failure investigator's recommendations; such reviews could be considered a hunt for maintenance-related failure causes.

11.2.4 Maintenance-related problems – (1.f)

Maintenance-related problems – (1.f) were involved because, as the sketchy records showed, a leaking cooling coil in the seal oil circuit had been found in one instance and the plant's mechanical workforce members had carried out quick repairs. On that occasion, no effort had been made to find the underlying failure cause. However, unless the root cause is known and addressed, a repeat failure is likely. It stands to reason that equipment owners would generally be well advised to establish the root causes of a failure and to implement long-term corrective action whenever possible.

11.2.5 Mechanical and assembly flaws – (1.d)

In this instance the plant's failure records also pointed to the possibility that, on at least one repair occasion, the HSLF's impeller had not been fully inserted;

consequently, the impeller hub probably did not make full contact with the shaft shoulder. The resulting impeller-diffuser contact damaged the impeller and may also have been responsible for a slight bend measured in the final output shaft.

Giving due consideration to repair procedures is of interest here. Prior to the final assembly a diligent maintenance technician will make it a practice to install a dial indicator. The indicator readings must ascertain that shaft runout is within acceptable limits. There should also be verification that the impeller nut has been re-torqued to the prescribed value. Since the axial length of the impeller hub will shrink as the previously heated impeller cools, re-torqueing should only be attempted after the impeller has returned to near room temperature. At that time, another indicator reading taken at the vane tips should confirm that impeller runout is not excessive. Again, maintenance must be carried out in accordance with step-by-step procedures.

11.2.6 Improper operation – (1.g)

Although there was no reason to attribute failure events to this cause category, the TFA wanted to highlight an important fact: Whenever machines have been designed and fabricated by reputable manufacturers, important data can usually be found in the manufacturer's *Operating and Maintenance Manual*. The data existed in this instance, "but who has time to read an entire manual?" Therefore, the TFA recommended that pertinent manufacturer's instructions be condensed and issued in single-sheet checklist format. Locations and items to check, dimensional listings and tolerances often fit on a single sheet of paper which, in turn, can be laminated in plastic. Single sheet checklists are greatly contributing to high rebuild quality and low failure rates at Best-in-Class (BiC) companies [2].

11.3 Mechanical seal upgrading

This owner-user company's maintenance technicians had shop-tested certain mechanical seals and found them to be marginal, at best. Therefore, the experienced TFA saw fit to question the appropriateness of using Teflon® wedge secondaries in this service and explained that more pliable Viton O-rings would be preferred as secondary sealing elements. In fact, as a dutybound and results-oriented investigator, the TFA proceeded to re-verify his recollection by telephoning a competent mechanical seal expert. The expert confirmed that his company had frequently retrofitted this style and model of HSLF compressor with seal upgrade kits, some at very moderate cost. The user-purchaser was asked to consider an upgrade kit and to communicate with two other mechanical seal manufacturers. He knew that HSLF compressors designed in the 1960s or 1970s will often benefit from recent advances in sealing technology. Such advances can involve alternative material compositions, seemingly minor configurational changes, improved flush plans, and advantageous seal water management systems [3].

Because the failure analyst was unable to rule out the (slight) probability of casing deflection due to pipe stress, he recommended that the machine's inlet and discharge flange bolts be removed during the next scheduled shutdown. Temporary dial indicators could be mounted, and casing movement monitored while the HSLF compressor pipes were being disconnected and reconnected.

During a debriefing meeting two operators recalled hearing a "chirping noise." It was thought that a gear tooth-related problem may have been causing the noise. A low-level noise originating in the speed increaser gear could possibly be managed, and the progress of gear tooth distress slowed down by using a synthetic oil blend made from polyalphaolefins and diester-based stocks. In HSLF compressor gears, modern non-foaming synthetic gear oils are superior to the automatic transmission fluids typically used decades ago. But regardless of circumstances and perceived sounds, it was again confirmed that compressor failure analysts or equipment troubleshooters will often have to deal with more than one deviation. Whenever these deviations combine, a distress or downtime event may need to be addressed with great urgency. This was the case here; it serves as a fitting reminder to use structured and repeatable approaches to failure analysis, and to take remedial action early.

11.4 Combining deviations in an automobile analogy

Finally, an automobile-related analogy and plausible scenario was used by the FTA to make an important point. He noted that we will occasionally be able to drive on unsafe tires. But suppose we're increasingly fond of our ability to run on sets of tires that are worn and underinflated. Our fondness probably lasts until the day when the pavement happens to be unusually hot, and we decide to place a heavy load in the car's trunk. That's when one of the two rear tires bursts, or "blows out." For just a few seconds the other rear tire manages to carry an extreme share of the already excessive load before it, too, fails.

We can be certain of how this chain of events will end; we also should agree that it's just not worth taking the chance in a personal car. Nevertheless, the risk of inviting far more costly catastrophic events is considerably greater on process machinery if we allow deviations to exist, accumulate, and become the "new normal."

11.5 Case Study No. 2 and the pitfalls of excessively high oil viscosity

This case study is again relating a consulting experience and resulting correspondence. It draws on an incident involving a pump-related fire; the pump-and-driver layout resembled Fig. 11.1 and details are given in this narrative. Initially, the sequence of events was investigated by refinery staff; a six-hour meeting was convened later. This six-hour close-out meeting was attended by a consulting engineer whose site notes and report later supplemented the observations offered by refinery staff.

11.6 Pump lubrication risks

In this instance, visualize a hot service pump which was lubricated with pure oil mist. Oil mist or "oil fog" consists of atomized oil suspended in clean instrument air [4]. The oil mist is carried to each bearing housing in a large header leading to small-bore tubing and "re-classifier" orifices that terminate at the bearings. The facility used ISO VG Grade 100 in its oil mist system, which is generally quite acceptable for the rolling element bearings in this 3,560 rpm, API-style pump. Therefore, some deviation from normal dimensions, or best-available application practice, or an external trigger event, must have existed to cause bearing failure. A process of elimination was employed to narrow the possibilities of what probably had happened.

The oil mist was properly applied between the bearing housing end caps and bearings; nevertheless, the angular contact back-to-back oriented thrust bearing set had failed. Calculations verified that the axial thrust load during normal operation was well within the bearings' allowable load range. At the time of failure, the various operating parameters had been quite normal.

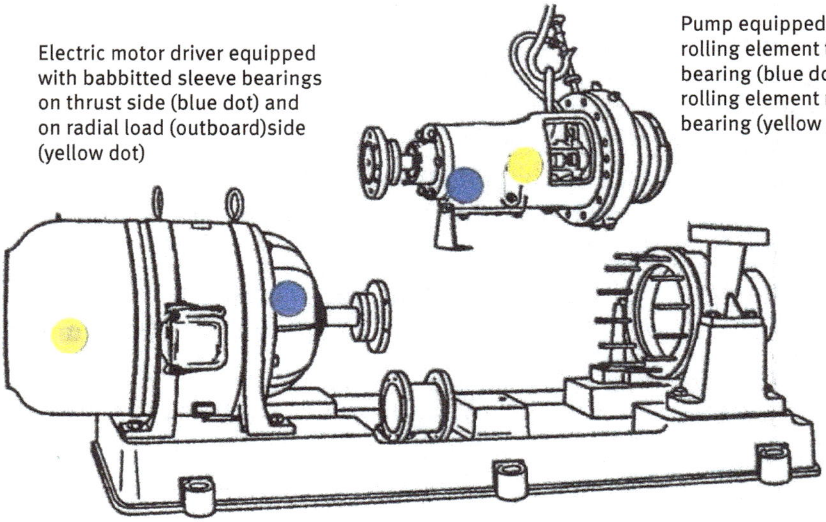

Fig. 11.1: Pump and Driver Layout Sketch showing bearing types, locations, and lubricants supplied (representation of Case 2 pump set).

11.7 How motor lubrication differed

The 115 or 150 kW/3560 rpm induction motor was equipped with babbitted sleeve bearings that carried the radially-acting rotor load. On its drive end ("DE"), the sleeve bearing had an integral thrust land that limited the rotor's axial travel. Lubricating oil was fed to each babbitted sleeve bearing by way of oil rings and constant level lubricators. It was noted that sleeve bearings are unusual on motors in this size and speed range; in general rolling element bearings are far more prevalent in motors up to 375 kW.

There is little doubt that ISO VG 32 was the required lubricant viscosity for these sleeve bearings. But the DE bearing had failed and a considerable amount of the low melting point babbitt had overheated and melted. It is well known that serious complications would result if one were to use – perhaps inadvertently – ISO VG 100 where ISO VG 32 is required. Oil rings will slow down in thick oil or, if the rings are not concentric within 0.002 inches (0.05 mm); chances are the authoritative Wilcock and Booser text [5] likely explains that it would probably stop moving altogether once it reaches values approaching 0.5 mm (0.020 inches). Moreover, a thick oil may not carry away enough heat to prevent babbitt from softening. With excessive heat buildup babbitt will quickly weaken to the point of melting.

11.8 Looking for deviations from acceptable operation

At the time of the close-out meeting at the refinery, none of the many attendees seemed to know if the (for babbitted bearings) excessively viscous ISO VG 100 had inadvertently been used. That the babbitt in the failed DE bearing had indeed overheated was not in dispute. A visible fatigue failure at the pipe thread-to-pipe juncture had allowed hot babbitt to flow out of the broken ¾-inch pipe which connected the constant level lubricator to the DE bearing housing. There was a puddle of molten babbitt on the concrete floor directly under the broken pipe; it prompted the consulting engineer to ask that an oil viscosity analysis be done on a sample that had been taken from the still-good, unaffected, non-drive end ("NDE") of the motor.

The oil ring retrieved from the distressed DE bearing housing was measured and observed to be 0.017 inches out-of-concentricity. This eccentricity is roughly eight times what ref. [5] allows. An eccentricity of 0.017 inches could have caused the flawed oil ring to go into a pendulum-like side-to-side motion. While this unintended combination of eccentricity and side-to-side slewing motion would not have caused catastrophic failure with *low VG* oil in the motor bearings, additional deviations increased the failure risk. But oil ring eccentricity combined with excessive oil thickness would have contributed to serious distress. In other words, failure in the DE bearing would result if the housing had been inadvertently filled with *thick ISO*

VG 100 lubricant. (If an oil analysis was indeed performed, the results were not made available to the consulting engineer.)

Considering a most probable time line, it was assumed that the pump thrust bearing had failed soon after the motor bearing had overheated, weakened, and begun to melt. Severe vibration originating at the motor DE would have caused radial overload in the pump thrust bearing and vibration-induced forces then led to failure of the threaded pipe.

11.9 Deviations from normal will add to failure risk

It is usually possible to operate machines with just one minor deviation. However, using a thick oil in an oil ring environment is an unacceptable deviation. Adding another deviation would have increased the failure risk even further. The consulting engineer believed there were four relevant deviations:

Deviation 1: As mentioned earlier, oil rings are almost exclusively designed for ISO VG 32 and VG 46 oils. (The specification sheet for the equipment at issue contained basic motor requirements but left open the properties of the oil). If the missing oil analysis proved that the oil was, in fact, VG 100, excessive oil viscosity would be the most probable sleeve bearing failure initiator. Only if it could be ascertained that VG 32 or 46 had been used would we have seen fit to delete lubricant issues from our failure inventory for the motor's DE bearing.

Deviation 2: The DE oil ring was found to be seriously out-of-round. Chances are that the oil rings in this motor had not been given stress-relieving heat treatment as an intermediate manufacturing step. The non-concentric DE oil ring likely slewed back-and-forth during operation and made frequent contact with stationary parts inside the bearing housing. Abrasion probably took place and caused the oil to be contaminated. Together with an insufficient amount of heat dissipation, the bearing was thereby compromised. Severe vibration resulted and caused the ³⁄₄-inch pipe to fail.

Deviation 3: Although an API-compliant flush-ground set of back-to-back mounted ball bearings was used, flush-ground bearings present an elevated failure risk in thrust-loaded pump bearings. If ISO VG 100 oil is used in conjunction with flush-ground bearings, the risk is somewhat elevated. Visualize that rolling elements in bearings arranged with the shaft horizontal (and most bearings in refinery equipment have horizontally oriented shafts) will have to plough through a small puddle of oil near the 6 o'clock position of the concave outer ring ball track. One of the two back-to-back mounted pump thrust bearings will then likely be unloaded because thrust (normally) acts in only one direction. The unloaded bearing will tend to skid, and the tendency to skid increases when viscous oils are used.

We can assume that other pumps at this refinery are similarly affected if flush-ground back-to-back sets of thrust bearings are used. While API compliance is usually optimal for general applications and for reasons of perceived cost savings, these bearing sets are not always the best choice; they will disappoint in many pump applications. Sets of pre-loaded and/or asymmetrical contact thrust bearing sets are easily cost-justified in situations where their use would avoid premature bearing failures.

Deviation 4: Smearing was noted in the machined "pockets" of the thick brass cages used in one of the failed pump thrust bearings. Whenever the unloaded half of such API-compliant duplex bearing sets (i.e., bearings with "pockets" machined into a thick brass cage) has one of its rolling elements pass through the small puddle of oil, that rolling element's surface velocity will slow down and, a fraction of a second later, its surface velocity will again increase. The difference in speeds results in skidding and the skid path is longer if ISO VG 100 oil makes up the small puddle. In bearings with thick bronze or brass cages (per API 610, but representing Deviation 4), skidding causes a transfer of material ("smearing") and some copper-bearing-alloy flakes from the many "smear islands" may fall into the ball track. Once that happens, failure can be instantaneous.

Among the potential reliability improving opportunities is the use of oil mist-lubricated *ball bearings* in motors up to 375 kW. Experience at Best-in-Class (BiC) refineries elsewhere has shown that oil mist lubricated ball bearings would save the user-purchaser considerable money over babbitted sleeve bearings and would provide an additional layer of protection.

11.10 When deviations combine, failure can be near-certain

There are several factors which acted together and increased the failure risk in this motor-driven pump set. The various influencing factors are described in many papers, articles, and books [2]. "Following the book" and implementing what others have successfully done years ago would be highly appropriate for this refinery. Above all, the facility should realize that we can often get away with one or two small deviations. However, allowing three, four, or even more deviations to combine will push fluid machines toward unpredictability and failure. It makes economic sense to identify and then disallow all risk-inducing deviations. Best available practice is quite simple: It allows no deviations. Whatever the deviation, Reliability Professionals should view its first manifestation as the event that triggers scheduled Inspection and Repair Downtime, occasionally abbreviated as "IRD."

What we have learned

- The Seven Root Cause approach to failure analysis is easy to learn; Best-in-Class companies are successfully using it at all job levels
- Allowing deviations to accumulate or combine will greatly increase failure risk
- Oil rings are usually designed for ISO VG 32 lubricants; they are prone to malfunction in thicker lubricants
- Sets of flush-ground rolling element bearings with thick brass cages tend to smear when allowed to skid. Skidding results from fully unloading a rolling element bearing
- Thicker oils can increase the risk of retardation of rolling element rotation, which describes skidding.

References

[1] Bloch, Heinz P. and F.K. Geitner. "Machinery Failure Analysis and Troubleshooting," 4th Edition (2012), Butterworth-Heinemann, an imprint of Elsevier, Oxford, UK, and Waltham, MA, USA; ISBN 978-0-12-386045-3.

[2] Bloch, Heinz P. and Allan R. Budris. "Pump User's Handbook – Life Extension," 4th Edition (2014), The Fairmont Press, Lilburn, GA 30047, ISBN 0-88273-720-8.

[3] Bloch, Heinz P. "Pump Wisdom – Problem Solving for Operators and Specialists," (2011), John Wiley & Sons, Hoboken, NJ, ISBN 978-1-118-04123-9.

[4] Bloch, Heinz P. "Petrochemical Machinery Insights," (2017), Elsevier, Inc., Oxford, UK, and Waltham, MA, USA; ISBN 978-0-12-809272-9.

[5] Wilcock, Donald F. and E. R. Booser. "Bearing Design and Application," (1957), McGraw-Hill Company, New York, NY.

Chapter 12
Common blower improvement opportunities

Blowers are low-pressure gas movers and many thousands are operating in virtually all industries. However, in the absence of rigorous industry standards or specifications, purchasers often make initial cost and rapid delivery their primary selection criteria. As a result of attractive pricing and de-emphasized long-term reliability, less-than-optimally-designed products tend to enter the supply chain. As just one example, they tend to purchase grease-lubricated equipment for inaccessible locations, such as the blower shown in Fig. 12.1. Many plant operators are later surprised that their equipment is maintenance-intensive, or unreliable, or both.

Fortunately, the weak links in simple blowers can be determined and remedied by thoughtful upgrading. Such upgrading includes well-executed failure analyses and cost justifications. As an obvious example, when a grease-lubricated pillow block bearing (similar to Fig. 12.1) failed several times within three years, someone decided to simply replace the failed bearing with one that had a much greater load capacity. Not much later, the owner-operator was confronted with the fact that this higher capacity bearing failed even more rapidly. Only then was it determined that the original bearing had probably failed because of a lubrication flaw and that the next three or four replacement bearings failed because they were too lightly loaded. Lightly loaded bearings tend to skid instead of roll, confirming the fallacy of "bigger is better" claims.

As explained earlier in Chapter 3: When the rolling elements in a bearing skid because they are too lightly loaded, the oil film is wiped off, and metal-to-metal contact takes place. Heat is being generated and degradation progresses at a fast pace. The facility went back to the originally supplied bearings and installed the small self-contained oil mist module shown earlier in Fig. 9.11. The bearings performed quite well and the belated addition of oil mist further moved the bearing life in the right direction. Using a synthetic lubricant of purpose-optimized viscosity added its own advantages and the earlier problems were left behind [1].

Along similar lines of reasoning, reliability-focused process companies today foresee blower situations where the addition of a dual mechanical seal cartridge (Fig. 12.2)drastically reduces gas losses. Used to provide both bearing lubricant and seal oil, a well-engineered circulating seal oil support system (shown earlier in Fig. 9.10) incorporates additional advantages over oil mist-lubricated bearings. A circulating seal oil support system can supply oils that do triple duty as a seal barrier fluid, coolant, and bearing lubricant [2]. This is but one of many examples where reliability improvements are well within the grasp of users and technology providers. User and provider join forces by investigating failure sources and origins. From there, they follow up by examining innovative upgrading opportunities and by calculating the cost justification for such upgrades. Combining the upgraded

https://doi.org/10.1515/9783110674156-012

Fig. 12.1: Blower pillow block bearing upgraded to oil mist lubrication (Source: Lubrication Systems LLC, Houston, TX).

Fig. 12.2: A dual cartridge seal can be considered a blower retrofit and upgrade component (Source: AESSEAL Inc., Rotherham, UK, and Rockford, TN).

aims of technology user and technology provider almost always achieves tangible progress. Collaboration encourages both to become value-adders.

12.1 Comparing aeration blower technologies

This text uses aeration blower technology as an example case for widely neglected opportunities to upgrade what we might collectively call "packaged systems." The author's aim is to convey the practical, experience-based approach for the owner's representative (the Reliability Professional) to take the lead in asset selection processes. In this instance, the entire aeration system is considered an asset.

12.1.1 Power consumer

As the main consumers of electricity in a wastewater treatment plant, aeration blowers can significantly affect the overall cost of plant operation. The most energy-efficient selection must be based on operating parameters, which are generally far from static. Moreover, energy efficiency is rarely the governing factor in optimized blower selection and comparisons made at only a single hypothetical operating point are inaccurate. We conferred with Aerzen USA (https://www.aerzen.com/en-us/applications/water-and-waste-water-treatment.html) (Coatesville, PA), one of the most experienced providers of high-reliability aeration blower technology, about modern selection criteria for air movers in this service category.

There is a clear need for up-to-date information that will help engineers understand the importance of more closely looking at the many factors that separate different designs. Selection and thoughtful integration of subsystems are of paramount importance, as is the overall operating philosophy of the owner-purchaser's facility. As of 2020, the Internet reference explained in more depth the extent to which subsystems and operating philosophies influence total cost of ownership.

12.1.2 Parameters that need to be compared

Experience shows that four parameters – site condition, type of aeration system, instantaneous mass flow of air, and related system variables both upstream and downstream – are important in blower selection. Available blower configurations are generally classified as either positive displacement (rotary lobe, Fig. 12.3), also dry screw (Fig. 12.4), or the dynamic (centrifugal) machines shown in Fig. 12.5 and 12.6. Dynamic blowers incorporate either a rotor with step-up gearing (Fig. 12.5) or a driver that can generate the requisite high (fixed or variable) speeds. Special purpose high-speed single-stage centrifugal blowers (Figs. 12.5 and 12.6) incorporate

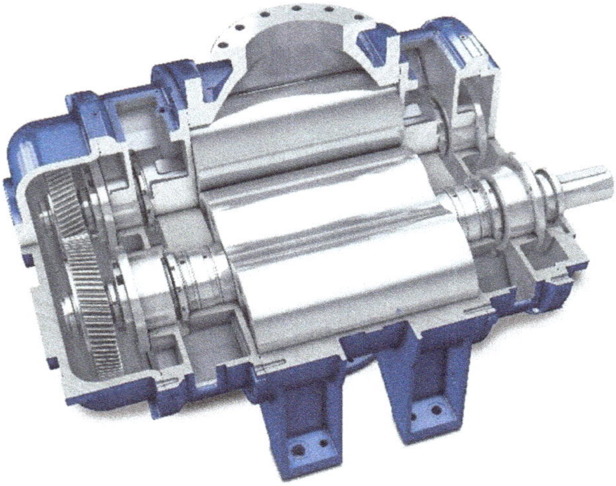

Fig. 12.3: Rotary lobe blower stage (Source: Aerzen USA).

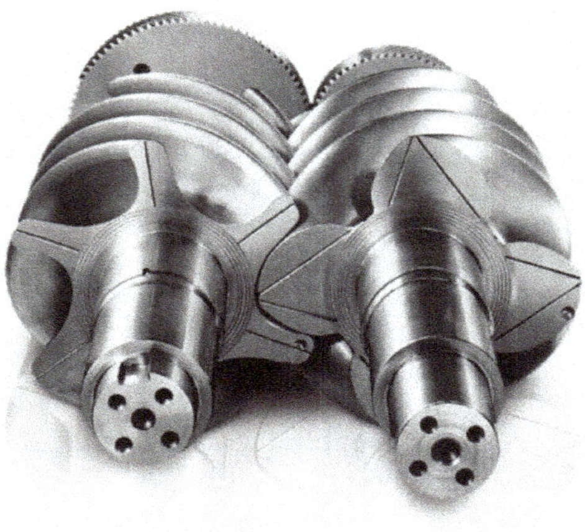

Fig. 12.4: Dry screw compressor rotors with timing gears (Source: Aerzen USA).

inlet guide vanes and outlet diffuser vanes to achieve wide turndown at constant operating speed.

High-speed single-stage centrifugal blowers can be operated with a VFD (Variable Frequency Drive) to provide additional flexibility while maintaining highest efficiency. They are also offered as self-contained packages, Figs.12.6 and 12.7.

A standardized high-speed single-stage turbo is depicted in Fig. 12.7. These packaged assemblies incorporate a three-dimensional radial-exit impeller, direct-

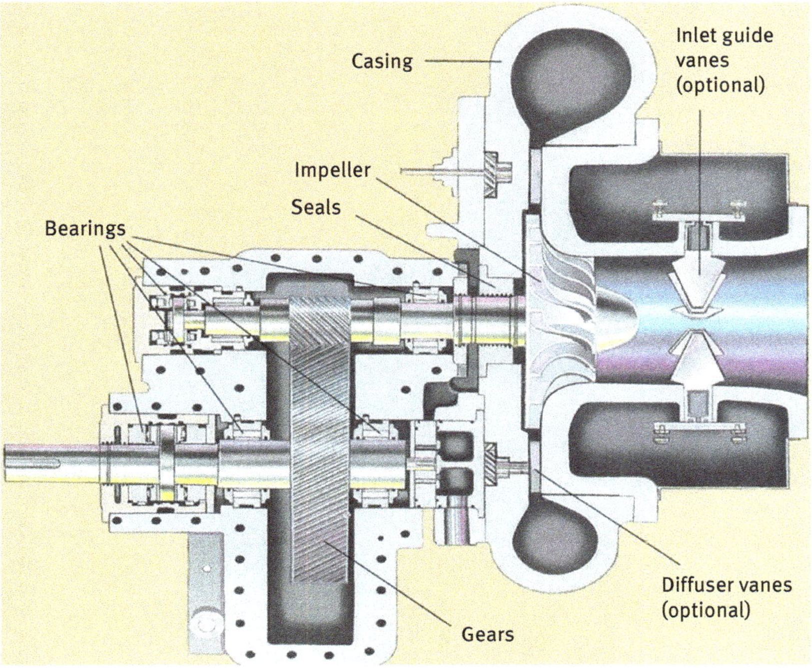

Fig. 12.5: Section of a centrifugal blower with inlet guide vanes and diffuser vanes (Source: Dresser Industries, Olean, NY).

Fig. 12.6: Single-stage centrifugal blower package (Source: Siemens Turbo).

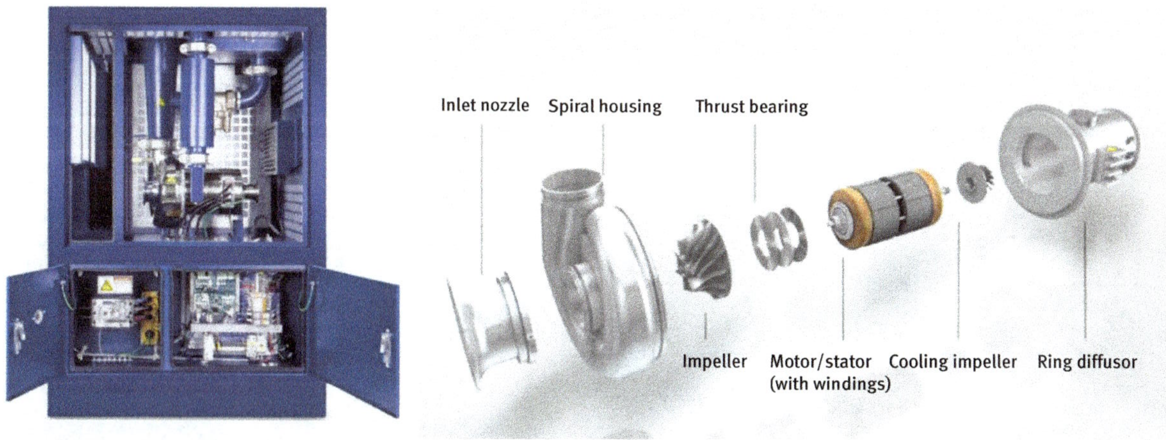

Fig. 12.7: High-Speed single stage blower package assembly (left) and exploded view of its turbo blower core with permanent magnet motor and airfoil bearings (right) [Source: Aerzen USA].

driven by a high-speed Permanent Magnet Synchronous Motor (PMSM). Combined with an integrated Variable Frequency Drive (VFD) and control panel, the package enables a wide operating range for continuous operation. Non-contact magnetic or air foil bearings are utilized in high speed turbo blowers to reduce friction, improve efficiency, and reduce maintenance.

12.1.3 Emphasis on modern technology

Aeration blower technology has advanced remarkably since the 1990s. The highly effective integration of control innovation and thoughtful component upgrading found in late-model aeration packages avoids downtime and increases the owner plant's profitability. Be sure to consider modern technology in 2020 and beyond.

What we have learned

- The selection approach must not take for granted that old technology is good enough
- Today's reliability professionals must not leave the field of aeration systems to the lowest cost provider
- Owner-purchasers should not allow EPCs (engineering and procurement contractors) to do piecemeal procurement and splicing together a system.
- The owner-purchaser's reliability professionals should be at the forefront of making relevant comparisons and then designating a provider who will accept turnkey responsibility for the entire wastewater aeration system

---

References

[1] Bloch, Heinz P. and Ken Bannister. "Practical Lubrication for Industrial Facilities," Third Edition, (2016), Fairmont Press, Lilburn, GA 30047, ISBN 0-88173-761-5.

[2] Bloch, Heinz P. "Petrochemical Machinery Insights," (2016), Elsevier Publishing, Oxford, UK, and Cambridge, MA, ISBN 978-0-12-809272-9.

Chapter 13
Integral gear CO_2 compressors offer high reliability and efficiency

In 1998, the experienced reliability professional mentioned in the foreword to this book engaged in a task requiring up-front compressor reliability engineering. Learning how he proceeded with his work gives us compelling insights on how a well-rounded user can, with a high degree of confidence, make an optimized selection of machinery. To the extent justified and appropriate, the optimized selection used by this reliability professional incorporated judicious extrapolations from prior experience.

His task started with researching suitable technology for compressing 2,684,000 m^3 per day (95 MMSCFD) of "bone dry" (–73.3 °C/–100 °F dew point) CO_2 from 1.15 barg (16.7 psia) to 187 barg (2,710 psia). The compressed CO_2 is delivered through a 330 km (205 mile) pipeline from a coal gasification plant in North Dakota to field operations in the Canadian province of Saskatchewan. The Canadian oil fields are among those that use the gas in enhanced oil recovery. As an added benefit to the Canadian environment, virtually all the injected CO_2 is expected to remain permanently sequestered in the depleted oil fields long after these will have been abandoned [1].

Typical of such initial research was, and also is, the investigation of available configurational options. Accordingly, three plausible compressor train layouts and casing selections were closely considered for this project:

(1) Motor + gear speed increaser + low pressure compressor + medium pressure compressor + high pressure compressor = five casings/bodies
(2) Motor + gear speed increaser + low pressure compressor + medium pressure compressor, and motor + pump = six casings/bodies
(3) Motor + integral gear compressor with 4 pinions (8 stages of compression), having a maximum pinion speed in excess of 26,000 rpm (a similar machine is shown in Fig. 13.1) = two casings/bodies

Implementation of Option 3 commenced within two years of initial scoping studies; the full results were available three years after fully commissioning Option 3. Two eight-stage machines were specified; a typical machine is shown in Fig. 13.2. Its eight impellers are depicted in Fig. 13.3.

The performance results could soon be fully evaluated and, as of 2020, system operation has been observed for almost two decades. The owner company in North Dakota has confirmed that integral gear compressors in dry CO_2 service achieved at least seven important benefits:

- Least capital cost for compressor purchase
- Reduced footprint, resulting in least installation cost

https://doi.org/10.1515/9783110674156-013

Fig. 13.1: A very large multi-stage (~50 MW) Integral Gear Compressor for air service in a PTA application (Source: MAN Energy Solutions SE, Oberhausen, Germany).

Fig. 13.2: An eight-stage Integral Gear Compressor for CO_2 service ready for final inspection (Source: MAN-Borsig, Berlin, Germany).

- Between-stage cooling and four optimum pinion speeds result in highest overall efficiency and lowest operating cost
- The chosen configuration represents the simplest mechanical design and yields high availability (uptime)

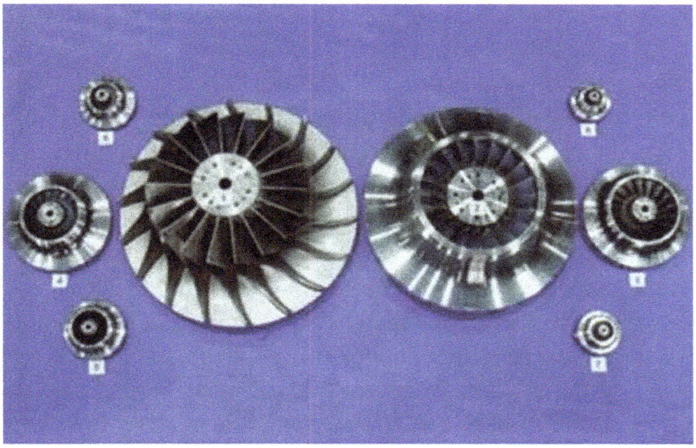

Fig. 13.3: The eight impellers before being installed in Fig. 13.2 (Source: MAN-Borsig, Berlin, Germany).

- Bull gear driven main lube oil pump simplifies the lube oil system by eliminating the need for an oil run-down tank in the event of loss of lube oil to the compressor
- Two 50% capacity 20,000 HP (~15 MW) machines can be selected to allow for partial continuous production in the event of one machine being down for maintenance
- Highest compressor flow turn-down has been achieved through incorporation of adjustable inlet guide vanes

Of course, manufacturers' capabilities, experience, and reputation were considered in selecting the compressor train supplier. The plant owners took important steps to ensure the success of this previously untested design. Budgetary and personnel resourcing were aimed at achieving high availability. With that in mind, a reliability design audit, life cycle cost analysis, and sub-supplier preference reviews were preplanned and carried out. During this comprehensive audit and review process (MQA), the shaft seals were given considerable attention and long-term reliability upgrades were made the focus of discussions between the potential manufacturer and the owner-purchaser. Agreement was reached on the final design; cost considerations drove the use of carbon ring seals on all compression stages. However, the seal housing dimensions would be large enough to accept dry gas seals in the (at the time uncertain) event that carbon ring seals proved unsatisfactory in the long range.

A factory performance test with CO_2 was performed to verify that the proposed aerodynamic and capacity performance was achieved. Of course, mechanical run test data were captured and closely analyzed.

Other mechanical features of these successful designs included state-of-the-art "Flexure Pad" bearings, oil-compatible O-rings, hydraulically fitted coupling hubs, and diaphragm-type spacer couplings at the driver connection (see Chapter 14). The motors were synchronous across-the-line starting designs with liquid cooling. An OEM-designed control system manages ramping up to speed and continuously monitors machine condition. The control system was integrated into the plant owner's digital control system (DCS), along with shaft vibration monitoring, determination of rotor thrust position, and bearing temperature acquisition – all arranged for automated alarm and shutdown when pre-defined upper limits are reached.

Within the control system, care had to be taken to ensure that no liquid would form in the various lower stage compression sections during startup and normal operation. Air-flooded inter-stage gas coolers are used. A glycol/water solution is circulated through the oil cooler and motor for cooling. Heat is rejected from the glycol/water solution through an air exchanger.

In 2006, an identical 3rd integral gear compressor was purchased by this facility to support additional sales of CO_2. The lubricant was upgraded from mineral-based oil to a synthesized hydrocarbon oil, and all three machines were upgraded at the same time. The change-over was prompted by the finding that mineral oil was being contaminated by certain sulfur constituents in the impure CO_2 stream. It was evident that the carbon ring seals allowed gas to leak into the lubricant. In keeping with the owner's reliability-focus, rigorous in-service testing, and comparison of the different oils under varying load and operating conditions were performed. A minimum efficiency gain of 2% was realized with the synthetic oil, in addition to the new oil having considerably higher resistance to degradation. The incremental cost of the synthetic oil was paid back within a few months. Moreover, the synthetic oil service interval was shown to exceed six years with no oil replacements needs in sight. The continued suitability of the synthetic oil is ascertained by periodic oil analysis.

Carbon ring seal life on high pressure stages 7 and 8 has met the manufacturer's warranty. However, the owner's plant experienced a measure of unscheduled downtime and recompression of higher-than-anticipated rates of CO_2; the gas quite obviously leaked back to compressor suction because of seal wear. The original seals fell short of the owner's goal of 20 years between seal repairs. Still, carbon ring seals for stages 1 through 6 have proven satisfactory. But the reliability professional's proactive insistence on allocating space for (future) dry gas seals proved fortuitous. In 2010 the gasification company, in partnership with a major manufacturer of dry gas seals and a group of engineers at MAN Turbo & Diesel SE, Berlin, Germany, initiated the design and development of dry gas seals (DGS) for this application. Dry gas seals were designed, built and factory-tested for the 7th and 8th stages and installation was planned whenever the next opportunity would present itself.

Overall, these eight-stage compressors have met and exceeded expectations and have resulted in cost savings of more than 10% compared to the initial investment needed to implement the next-best option. The machines demonstrably had the lowest operating cost and achieved at least 96% availability. With increased interest in CO_2 sequestration projects, enhanced oil recovery and urea production, this technology clearly merits consideration.

But the main lesson is in observing how a motivated owner's engineer took the lead in persuading one of the world's most prominent compressor manufacturers to do rigorous and uncompromisingly diligent engineering. This reliability engineer became a value-adder by first studying and understanding advanced improvement options. He was then able to "sell" these options to the leading compressor manufacturer. We may conclude that he proved that nothing is impossible for informed reliability professionals who strive to implement action based strictly on knowledge. Reliability professionals make it their business to disseminate facts and seek out open-minded manufacturers whose spirit of cooperation and innovation turns both parties into firm allies. We can be sure that the allied partners mutually benefited and to this day continue to benefit from the approach described here.

What we have learned

- Talented professionals are valuable idea generators
- Best available technology can either be implemented or planned for before the risk-averse and indifferent get moving
- Better couplings (see Chapter 14), Flex-Pad bearings, synthetic oils, O-ring materials chosen with future lubricants in mind, advanced dry gas seals improve even the most modern gas compressors
- Manufacturers' capabilities, experience, and reputation must be considered in selecting process compressors
- Budgetary and personnel resourcing must be aimed at achieving high reliability and availability
- Reliability design audits, life cycle cost analysis, and sub-supplier preference reviews are up-front budgetary items
- During audit processes, shaft seals and other relative unknowns must be given considerable attention and long-term reliability upgrades made the focus of discussions between potential manufacturer and owner-purchaser

---

Reference

[1] Robert DeMaria. "Consider Integral-Gear Compressors in CO_2 Services," *Hydrocarbon Processing*, May 2013. Also published in Turbomachinery International, July 2016.

Chapter 14
Power transmission couplings: select the best configurations

The American Petroleum Institute's API 671 Specification describes in detail the requirements for "Special Purpose Couplings for Petroleum, Chemical, and Gas Industry Services." The specification is primarily intended for equipment that is normally non-spared and is critical to the continued operation of the installation. This could be a 200 kW, 3,600 rpm motor-driven centrifugal pump or a 60,000 kW, 11,000 rpm turbine-driven compressor. The fluid machinery owner-user normally determines which machines are critical to successful plant operation.

14.1 Historical developments

Through most of the twentieth century, gear couplings (grease-packed or oil spray-lubricated) were considered the best choice for critical machines. Gear couplings were relatively simple to design and manufacture. They were generally reliable and readily adapted or extrapolated, even as machine rotors increased in span and speed. However, as machines became more reliable, the weak links or impediments to increased equipment run length became more obvious, and the main disadvantages of gear coupling became better known:

- The sliding action of the gear teeth, even under normal operating conditions, produced high reaction forces at the machine bearings.
- Effective operation required some radial clearance at the gear mesh, limiting concentricity, balance quality, and repeatability of balance.
- Lock-up of the gear mesh, in extreme cases, could cause catastrophic bearing damage and considerable downtime.

All the cost savings initially realized with a gear coupling, even after 5 or 10 years of successful operation, will vanish if excessive vibration or bearing damage due to lock-up cause an unscheduled outage. That is where non-lubricated, "dry" couplings are of importance.

14.1.1 How "dry" metallic couplings function

Diaphragm couplings carry torque as a shear stress between outer and inner diameters. Couplings by the Bendix Corporation led the way (Fig. 14.1). Multiple discs

https://doi.org/10.1515/9783110674156-014

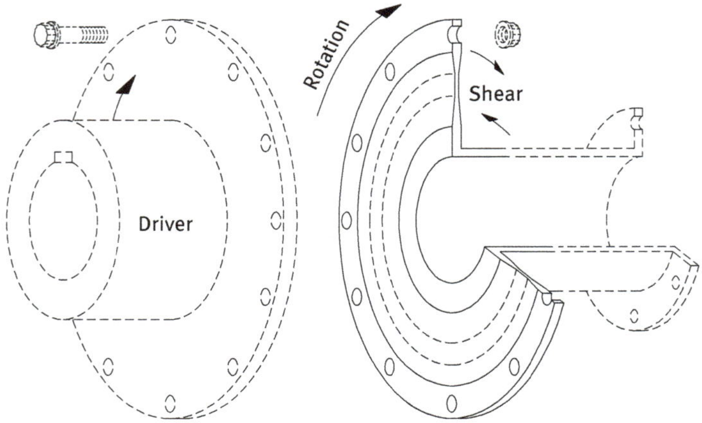

Fig. 14.1: Single diaphragm coupling developed by Bendix, early 1960s (Illustration courtesy of David Matt, FlexElement Texas Inc., Houston, TX).

(Fig. 14.2) were soon offered by Bendix, Zurn, and others; a special version is still marketed by Coupling Corporation of America (CCA).

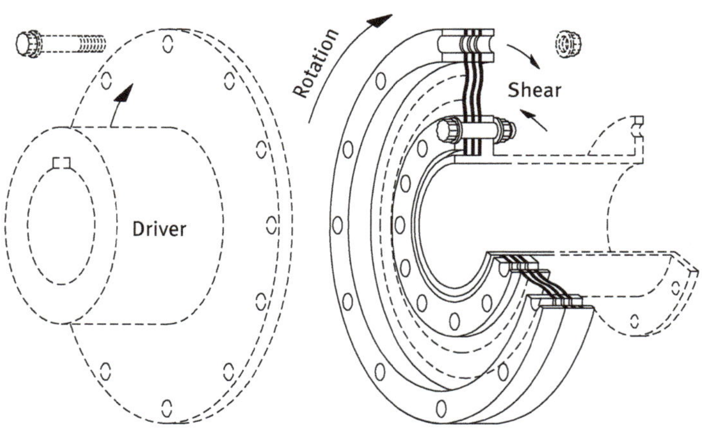

Fig. 14.2: Multiple diaphragm coupling developed in the late 1960s (Illustration courtesy of David Matt, FlexElement Texas Inc., Houston, TX).

All disc couplings carry torque load as a tensile stress in the tangential link (Fig. 14.3). Steel is typically twice as strong in tension as compared to operation in shear. Accordingly and for a given torque capability, the disc coupling's outer diameter is generally lowest overall. In terms of flexibility, disc couplings are stiff in the two directions where stiffness is needed (torsionally and radially); they are soft in the directions

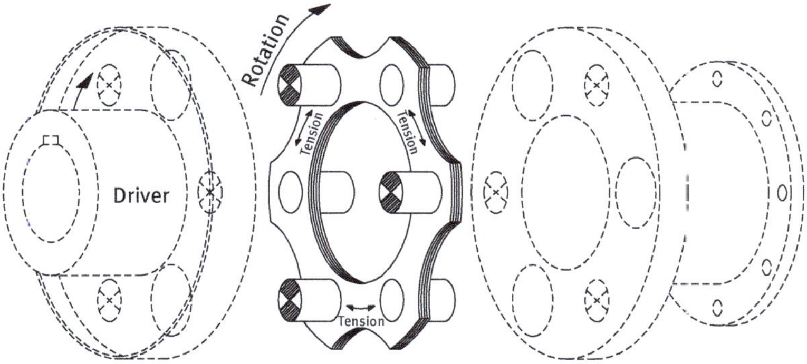

Fig. 14.3: Typical disc pack couplings underwent refinement since the mid-1960s (Illustration courtesy of David Matt, FlexElement Texas Inc., Houston, TX).

where they need to be soft, i.e., angularly and axially. This represents a favorable combination of mass-elastic characteristics, which also suggests that both overall rate of acceptance and range of application for these couplings will continue to expand in the reliability-focused environment of 2020 and beyond.

Looking back at history: With time, there developed a need for light weight, low-moment couplings, and good solutions quickly presented themselves. Beginning in the 1960s, the Bendix single-diaphragm flexible element coupling became the first non-lubricated coupling to gain widespread acceptance with original equipment manufacturers (OEMs) and machinery users alike. In the 1970s, the Zurn Company produced a valuable variation on the diaphragm flexible element in the form of a multiple-diaphragm coupling (Fig. 14.2) which offered a significant reduction in stiffness (both axial and angular) when compared to its single-element predecessors. Against this background, both single and multiple-element diaphragm couplings have been used for many years on high-energy turbomachinery and have been proven capable of many years of safe, reliable, highly predictable operation.

Beginning in 1980, disc-type couplings (Fig. 14.3) have become an increasingly popular choice for machinery users and OEMs. Originally manufactured under the Thomas brand name since the late 1920s, disc couplings by Turboflex and others have been in use from the late 1930s, although for many years these couplings were considered not suitable for high performance applications. However, refinements in component design (optimized by using Finite Element Analysis), as well as the introduction of PTFE (Teflon®)-coated discs, have resulted in disc couplings that offered all of the advantages of the single or multiple-diaphragm coupling together with the added advantages of a minimum outer diameter and relatively large inner diameter. These attributes made disc couplings ideally suited for low-moment applications.

14.1.2 Excellent life expectancy

Modern non-lubricated flexible couplings using either metal diaphragms or metal discs are capable of absorbing stresses due to misalignment, both angular and parallel. When properly selected, they will function for many years in continuous operation without the need for inspection, maintenance, repair, or replacement. It would be as inappropriate to consider non-lubricated metal flex part couplings a wear-prone component as it would be to so label the main rotor shaft. Both components can be designed to absorb high levels of combined steady state and cyclic stresses, and each can have a useful life approaching infinity. Careful selection and application of couplings for use on high energy machinery is possible; proper selection will result in years of safe, reliable operation. There will be no need for scheduled replacement or stand-still maintenance; moreover, simple and infrequent inspection procedures can be followed. By using a stroboscopic light on relatively small machines, it is often possible to examine the extent to which a "disc sandwich" has deformed or is flexing while the equipment operates.

14.1.3 Initial choice

The initial coupling choice must be based on its torque capacity (using an appropriate Application Factor or Factor of Safety) and linked to its misalignment capability. A system-based review is conducted to carefully examine the mass/elastic characteristics (weight, inertia, center of gravity, axial, angular, and torsional stiffness). This review will ascertain that adequate, safe margins are maintained between operating speed and prevailing lateral, torsional, or, in isolated cases, axial resonant frequencies. However, disc couplings with non-linear spring stiffness characteristics are not susceptible to axial resonance.

Rigorous Rotor Dynamics Analysis (RDA) techniques have been developed, and programs are now widely available which, when properly applied, permit a detailed assessment of the coupled rotor system long before the design is finalized and before manufacturing commences. These same techniques, in the hands of expert machine design engineers, can be used to ensure safe, successful coupling retrofits.

In other words, performing an RDA study is highly recommended and is one of the keys to long-term reliability of major turbomachines. There have been instances where false economy led to excessive risk taking by machinery owners. The result, often preceded by years of poor performance and high vibration, proved the original choice of coupling to have been bad. With couplings sometimes dangerously mismatched to either rotor geometry and/or operating conditions, there have been near-misses and expensive downtime events. Reliability-focused owner-purchasers and especially those who optimally use the MQA concept, report never having experienced issues with non-lubricated couplings.

14.2 Comparison of coupling flexible elements

Gear couplings continue to be occasionally retrofitted to older fluid machines; another common approach is to replicate the geometry of the gear coupling so that the "dry" coupling fits in the originally allocated space without major guard modifications. The mass-elastic (meaning weight, center-of-gravity, overhung moment, WR^2, torsional stiffness) characteristics are often duplicated to avoid changes in torsional or lateral critical speeds.

The approach described in the previous paragraph assumes that adequate margins exist – with the original coupling in place – between running speed and lateral, torsional, or axial resonant frequencies. However, a rigorous RDA (Rotor Dynamics Analysis) should always be considered as an appropriate means, as smart insurance, to make an informed choice of coupling when contemplating retrofits on high speed machinery.

14.2.1 Torsional stiffness

Because connected machines function as spring systems in series with another, the potential exists for torsional resonance. Couplings, being the more accessible components within the system, can be, and often are, modified to "tune" the overall system, such that potentially dangerous torsional vibrations are attenuated or avoided. Both the flexible element configuration and/or, more commonly, a coupling spacer section can be redesigned to produce desired changes in total torsional stiffness.

The coupling spacer sections in Figs. 14.4 and 14.5 depict common hybrids. Figure 14.4 shows a typical spacer section designed either to increase torsional stiffness or to increase lateral stiffness and increase whirling speed on long spans such as cooling tower fan drives. Figure 14.5 shows a quill-shaft spacer section designed to optimize torsional stiffness. An unlimited number of iterations can be made to combine specific spacer diameters and wall thicknesses. Of course, these are governed by strength and service factor (Factor of Safety) considerations. Accurate

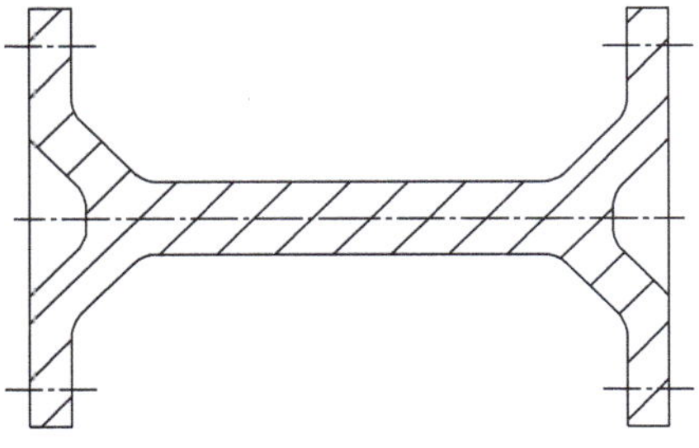

Fig. 14.4: A typical spacer section designed either to increase torsional stiffness or to increase lateral stiffness.

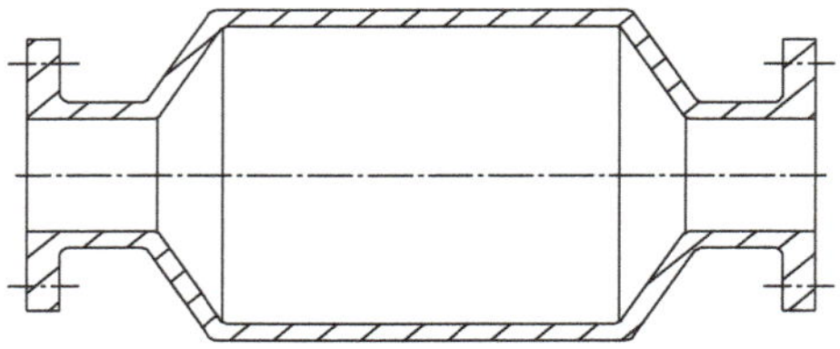

Fig. 14.5: A quill-shaft spacer section designed to optimize torsional stiffness.

computer predictions are entirely feasible; these are employed to avoid operating at or too close to resonant conditions. Operation at resonance would cause stress amplification.

Nevertheless, torsional "criticals" can be of interest in many ways. In at least one retrofit application, the proposed non-lubricated coupling was shown, using RDA, to provide an adequate safety margin between operating speed and torsional "critical." However, the same RDA demonstrated that the original gear coupling had been in successful service for over 20 years, operating only a single percent removed from potentially damaging torsional resonance. Not having a gear unit in this syngas train had proved fortuitous; it had greatly reduced the probability of encountering torsional excitation.

The owner-purchaser had asked one of their in-house SMEs (Subject Matter Experts) to coordinate project implementation. The SME obtained valuable RDA updates while tracking the work progress and ascertaining the appropriateness of the various analysis steps.

What we have learned

- With very few exceptions, non-lubricated couplings are ideally suited for fluid machinery in virtually all sizes, torque transmission requirements, and operating speeds.
- Both original equipment manufacturers' scopes of supply as well as retrofits and upgrades typically involving experienced coupling design and manufacturing companies should be considered.
- The use of these couplings should be maximized in fluid machines.
- Properly designed and manufactured (extrapolations are possible, but engineering analyses are always appropriate), these couplings have unlimited life.

Appendix 1
From early condition monitoring to wireless sensor technology

We asked Dr. Ian Newby for expert input on a subject that clearly affects our overall understanding of fluid machine reliability and mechanical performance. As an authority on the subject, he agreed to provide material of interest and allowed us to condense some of his writings. Better yet, he encouraged us to continue the discourse at ian.newby@aesseal.com.

A.1.1 Background and history

Condition monitoring started to be used in general industry in the 1970s with main frame-based systems. In the late 1980s the systems became PC-based and dedicated physical data collectors were widely used. The data collector approach moved across to Microsoft Windows in the mid-1990s. As of 2020, most systems still use this model, with a small number of providers favoring fully web-based systems.

The hand-held data collectors used with main frame and PC-based systems were traditionally custom designed hardware, with dedicated electronics for data acquisition. The displays were often small, and the data collectors were operated by a limited number of hardware buttons located on either side of the screen. It should be noted that data *collection and acquisition* are terms used for an activity that precedes data *analysis*. Data analysis dissects the monitoring task and allows an analyst to determine which component part is defective and whether the defect is linked to a machine speed or range of speeds. Data analysis broadens our knowledge of how and why parts fail. In turn, understanding the possible contributors opens the way to specifying remedial action.

At a few top user companies (generally called "Best-in-Class Performers" or "BiCs"), data collection was (and still is) viewed as the activity which gets the *person* close to the machine. Suppose one takes weekly vibration readings on a fluid machine that normally lives an average of seven years (or 7 × 52.2 = 365 weeks). It could then be assumed that 364 data taking events uncover no distress. However, we certainly would not feel comfortable employing this reasoning, and in the span of 364 weeks never look for high vibration. So, what reasoning *do* we see employed at BiCs? It is this: We want an operator to leave the comfort of the control room; we expect our operators to act as the eyes, ears, and nose of a plant. We likely believe that it will take time to devise sensors and artificial intelligence to run a complex process plant without operators. But chances are that for every single vibration excursion found, an operator has *seen* a seal leak, *heard* a hissing sound from a

https://doi.org/10.1515/9783110674156-015

defective steam line gasket, *smelled* a process gas escaping from a compressor piston rod, or whatever. For now, though, we might defer this conversation and steer it back to sensor history.

A.1.2 Highlights of current usage

Smart phones. With the advent of smart phones and tablets some manufacturers have designed data collectors where the user interface runs on these readily available devices. The collector is responsible for data acquisition, and for transferring the data collected to the phone or tablet wirelessly or via a physical connection. This allows far simpler data collector hardware to be designed; the benefits to user and manufacturer are more rapid development and reduced costs. The sped-up improvement of commodity phones and tablets also means that the applications are getting faster every year as they benefit from the new devices reaching the market. Phones and tablets are available in various form factors and protection levels, with industrial, HAZLOC (a classification system for hazardous locations), and ATEX (an abbreviation of ATmosphere EXplosible, French language) devices available from several manufacturers.

As mentioned above, vibration analysis is one of the most common techniques for condition monitoring of rotating equipment. Over the five or so years leading up to 2020, this area of endeavor has seen the release of a wide range of low-cost wireless sensors, potentially making condition monitoring available to a whole new range of assets. Traditional hand-held vibration analysis systems are expensive, costing well in excess of $20,000 per data collector, with additional outlays required for software licensing, annual support, and infrastructure.

Hard-wired systems are also available. These are typically fitted in close proximity to the assets to be monitored and can collect data from multiple sensors (typically up to 32 different channels can be connected). These systems tend to be expensive both in terms of hardware and installation costs; they usually need access to power and network connectivity. Typical costs for this type of system start around $1,500 per channel, including configuration and installation.

A.1.3 Overview of vibration analysis systems

Vibration analysis systems can be split into three categories based on the type of data they store. This will be our brief overview.

A.1.3.1 Full time waveform collection

Systems that collect the entire time waveform have the advantage that they store the raw signal. Time waveforms can be post-processed by the system to generate both the vibration spectra and extract other parameters for trending.

This approach allows historical data to be viewed in many ways, and new trends can be created from the historical data. It can also be quicker, as a single reading can be taken and post- processed to produce spectra and other parameters. Time waveforms have become more accessible since device processing power has increased. Today, in 2020, signal processing can be performed easily and quickly in software on the collection device rather than by dedicated electronics, as was the case in early data collectors. Full time waveform collection has the largest data storage and transfer requirements, with a single reading able to exceed 32 KB.

A.1.3.2 Spectra collection

Historically, data collectors have had limited on-board storage capability, which limited the size and amount of time waveforms that could be collected. For this reason, many systems default to only storing a spectrum view even today, which allows a minimum of a two fold increase in the number of readings that can be collected for the same memory. Spectral averaging was often used to allow a longer time signal to be measured with a much smaller storage and transfer requirement, but with a loss of spectral resolution. This approach needs less storage (at most ½ as much) but is not as flexible for analysis. Often multiple readings will need to be collected at the same location (especially if a time waveform and spectra are required). This results in more time spent collecting data.

A.1.3.3 Parameter extraction

The final method of storing data is to process the signal on the device, extract the desired parameters from the raw waveform or spectra, and then store only data of interest. This has a massive impact on the volume of data which are stored; storage, and data transfer requirements can be over 1,000 times less than full waveform collection.

This approach is often used for displays somewhat analogous to traffic light displays. The light can tell you there may be an issue, but it is incapable of telling you what the issue is. Additional readings taken with a more elaborate data collector are then required to perform the analysis and determine what may have to be interrogated to determine an actual failure mode. Again, an adjunct to the traffic light can inform us how long it takes to get to Houston Intercontinental Airport and what detours to take ….

A.1.4 The evolution of mobile hand-held vibration analyzers

The first generation of mobile vibration analyzers typically consisted of bulky devices, with custom electronics and proprietary software. By the standards of modern smart devices, these early analyzers had ridiculously low specifications, 8 MB storage, 4 inches–5 inches displays and limited physical buttons were commonplace. Due to the development cost of these devices, they were usually available on the market for many years with only minor improvements, among them memory capacity increases incorporated on the newer versions. Manufacturers typically developed new models every eight years or so, but intrinsically safe models were often kept on the market far longer due to the increased development costs and low sales volumes.

As of 2020, all the traditional data collector manufacturers were still using this same development cycle. However, an alternative approach taken by some manufacturers was to base their systems on commodity hardware. Early examples used Windows CE Personal Digital Assistants (PDAs) with additional plug in modules for the data acquisition portion built as expansion modules. These systems were much simpler and quicker to develop; their electronics were configured only for data acquisition. There was no need to provide user interface hardware or storage; these capabilities resided in the PDA.

A second generation of this approach materialized when the data acquisition hardware became independent from the PDA. This independence is linked to the development of rechargeable data collectors controlled via Bluetooth from the host device. From then on, any PDA running a supported operating system could be used as a vibration data collector. Included were specialist devices with different form factors, and additional features such as Barcode/RFID Readers, ATEX approval, and 3G connectivity.

With the demise of Windows CE as a platform, most of these systems have moved across to Android or Apple as their platform of choice. Android, in particular, has a large number of devices available with built-in industrial features such as barcode readers, NFC readers, ATEX approval and thermal cameras. Using commodity hardware also means the applications run faster every year as release cycles are much faster. A traditional vibration data collector costing $20,000 and up could thereafter be replaced by an Android or iOS-based system for around $2,000, with correspondingly lower costs for software licensing and annual support. A full web portal can be used to manage, analyze, and report on the data collected.

A.1.4.1 Accelerometer technology

Accelerometers are the most common type of sensor used for condition monitoring. Traditionally, these sensors are connected to and powered by the vibration analyzer via a short, rugged cable. Accelerometers are historically piezo-electric in nature; they typically allow measurements to be taken up to 20 kHz.

Tri-axial accelerometers have also been routinely available and used with higher-end traditional data collectors (multiple channel acquisition capability is required). Typically, these are far larger and more expensive than using three single axis accelerometers. The most recent breakthrough came with the development of MEMS (Micro-Electro-Mechanical Systems) accelerometers. These are a fraction of the size and require a fraction of the power compared to traditional accelerometers and are one of the key enabling technologies behind the new generation of wireless sensors. Different MEMS accelerometers have different frequency and amplitude ranges (such as 8 Khz with a 20 g range) so care must be taken when choosing a MEMS sensor for use. MEMS accelerometers often have a weaker third axis due to their construction so this should be taken into account.

A.1.4.2 Labor requirements

Hand-held devices are carried by technicians to the assets to be monitored. The sensors are placed on the location to be measured and a reading is taken. Typically, three readings are taken (horizontally, vertically, and axially) as close to each bearing on the asset as possible. These readings are usually reviewed later when the data are uploaded to the software suite, although some analysis takes place on the device.

Traditionally, data collection involves sending a technician to every machine to be monitored to collect readings. With early data collectors this was a very time-consuming process with each measurement taking minutes to complete. A modern system utilizing tri-axial time waveform collection can take three readings in perhaps 10 seconds. One site which migrated from an early system in 2017 used to require five days just to collect and analyze the data. When they upgraded to a modern system it took a day to collect and analyze the data, doing so with greater accuracy.

In about 2008, a good rule of thumb assumed that an engineer could collect and analyze data for 10 assets per working day. With the new data collectors based on commodity hardware this has increased to 65 assets per day. Most of this gain is attributable to minimizing time spent per measurement location. Traditional systems required multiple measurements per measurement location and direction, with separate measurements taken for the spectra and the time waveform. A fully tri-axial system just taking time waveforms will reduce this by a factor of 6.

A.1.4.3 Wireless sensors

The breakthrough in recent years has been the availability of high-quality wireless sensors with significant on-board memory and processing power, which along with the mass adoption of smart phones and tablets greatly simplifies their installation and use.

Memory technology has improved rapidly and reduced massively in cost since the turn of the century. Full-time waveform data collection is now easily achievable – 50 MB of storage (small by the standards of today) can store over 1,500 typical time waveforms, which is more than could be collected in in a single shift by a traditional hand held data collector. For wireless sensors, then, memory capacity can no longer be considered an excuse to not perform full-time waveform collection. The big issue with full-time waveform collection therefore is data transfer speed.

Most sensors allow some method of running detection. Running detection can discontinue the process of logging when the asset is not running. This, quite obviously, reduces the volume of data to be transferred.

Wireless sensors can be placed in three main categories:

- **Low energy wireless.** Low energy wireless systems (including Bluetooth Low Energy, ZigBee, and Wireless Heart) are excellent at transmitting very small amounts of data for a very low power requirement. Most of the wireless sensors on the market today use one of these technologies. The main disadvantage of these is low speed when transferring anything other than tiny amounts of data – a typical time waveform could take up to a minute to transfer using low energy wireless. The standby power requirements are very low, allowing the sensor to be connectable, at all times. Most of these systems will only transfer parameters extracted from the measurement, or at best a very limited number of low-resolution spectra. This creates issues whether the data is being manually collected or automatically collected via a gateway. If manually collected, the user will have to wait at the asset for a long period of time while readings are downloaded from the sensor. For automatic collection this limits how many sensors can go via the same gateway for a given measurement interval, for hourly collection with tri-axial sensors a gateway can only connect a maximum of 20 sensors before becoming overwhelmed if full-time waveforms are collected.
- **Wi-Fi/Bluetooth Classic.** Many sensors that transfer via full Wi-Fi or Bluetooth Classic are available in the market. These can transmit much faster than the low energy versions but draw much more power. Over five full-time waveforms can be transferred per second using this type of technology, but battery life is much reduced as their power requirements are much higher. Some devices use a physical button on the sensor to turn on the Wi-Fi in an attempt to increase the battery life – unfortunately this means that physical access to the sensor is needed to download data, defeating one of the main reasons to install a wireless sensor.
- **Hybrid devices.** Hybrid devices use both low energy wireless and full power together to get the best of both worlds – use the low power connection to switch the device on then use the full power connection to transfer the data. This gives

good battery life and great transfer speed. These devices can allow the full-time waveforms to be collected while retaining good battery life, allowing full fault diagnosis to be performed from the collected data.

A hybrid device will give by far the best balance between battery life and usability: allowing the ability to collect full-time waveforms for fault diagnosis while minimizing the time it takes for data transfer to be completed.

A.1.4.4 Wireless sensors: Collection strategies

There are two main methods of acquiring the data collected by wireless sensors: manual and automatic.

Manual collection involves someone visiting each sensor and collecting the data using either a custom data collector or by an off the shelf device, such as a smart phone or tablet. Manual collection allows other checks to be performed on the asset, such as a visual inspection or lubrication while the data are being collected. This can be a very low capital cost, but potentially high labor cost.

Automatic collection is a more expensive option, requiring at least a single gateway which polls the sensors and uploads the data to the Internet (typically using 3G/4G technology). Depending on the installation, a large number of gateways could be required due to the range of the sensors. To alleviate this, some systems include repeaters, which reduce the physical number of gateways required. Most gateways and repeaters require mains power – the few battery powered repeaters on the market suffer from poor battery life as they have to be awake for longer to repeat signals from multiple sensors back to the gateway. This approach has a very high capital cost, but low direct labor cost.

A.1.5 Upgrading to new wireless sensor technology

A.1.5.1 Economics of wireless sensors

The current cost of wireless sensor hardware suggests that they are often not suitable for every asset in a plant; a manual (hand-held) condition monitoring system can cover most of the assets in a plant effectively for lower cost and forces the operator, our "eyes, ears, and nose of the plant" to use these senses at or very close to the asset. For critical systems, a hardwired solution is often desirable which would be able to shut the asset down if a fault is detected.

We might now ask: Where is the sweet spot for wireless sensors? Assets which are run infrequently (but often missed when a hand-held route is chosen) and assets which are difficult to access due to guards or location, are the best candidates for the wireless sensors on the market in 2020. Other uses include equipment such as air handlers which may be theoretically accessible, but the very act of opening the cabinet to gain access can affect the running of the equipment and result in readings which do not reflect the true operating condition. In addition, wireless sensors can be used to successfully capture sudden failure modes because they can monitor much more frequently than the traditional far less frequent (perhaps monthly) reading.

Typically, vibration readings are collected monthly (the most prevalent failure modes are typically detectable over three months in advance). A good rule of thumb is that 65 assets per day can be monitored by one person, equivalent to 1,300–1,400 assets per month. For the same level of coverage, one would typically use three sensors per asset, equating to 3,900 wireless sensors required. Cost wise, one would end up spending over 10-times the capital cost on current (2020) wireless sensors to replace the collection coverage of a single hand-held data collector. It follows that wireless sensors are not currently the final answer to low cost condition monitoring. For the time being, sensor selection should be considered part of a mixed solution which includes hand-held data collection and often hard wired systems on critical assets.

A.1.5.2 Wireless sensors: data storage and access

Wireless sensors can generate enormous amounts of data which must be downloaded, stored, and analyzed. How accessible the data are once it has been collected is a key factor to take into account when choosing a sensor system. Some systems require software to be installed on dedicated hardware which is installed within the site. The data are then aggregated to this dedicated hardware. Systems which require dedicated client software installed on end users' computers are an additional complication which can be a barrier to successful deployment. The usability and flexibly of the software suite should also be taken into account when choosing a sensor system.

The easiest systems to install connect directly to the Internet and are accessible via a browser. No server or client software configuration needs to be performed and no dedicated server hardware needs to be sourced. The browser-based software allows easy access from anywhere with an Internet connection.

The temptation to always configure sensors to collect data very frequently should be discouraged. Most failure modes take months to go from first detection to causing an issue on the asset; accordingly, too frequent collection is simply a waste of time (both in collecting the data and reviewing it) and also a waste of battery life. Most sensors can be configured with customizable log frequencies ranging from minutes to days. In most cases daily or even weekly readings would be more than acceptable. There are reasons to log more frequently: For troubleshooting purposes where process variations may cause faults to appear infrequently, creating a baseline sequence of readings on newly installed or repaired equipment, or where lead time to failure is so short that automatic shut-off of the equipment is required.

A.1.5.3 Wireless sensors: why implement wireless sensors now?

Many systems have identified problem assets by comparing readings to a single set of baseline readings collected when the sensor was first fitted. This causes problems as process variation can massively affect the level of readings. Wireless sensors can collect many more readings after installation while processes are running, all in a shorter timeframe than traditional hand-held collection. By utilizing statistical alarms on the readings collected from wireless sensors for a period after installation, alarm and warning levels can be configured appropriate to the individual asset allowing asset degradation to be readily identified from the trends of the readings collected.

Once a problem is identified it is necessary to identify the underlying cause. The new generations of wireless sensors store the fulltime waveforms which can be used for full fault finding without the need for additional data collection by a more fully featured data collector. This is a hugely important feature.

Most of the wireless vibration sensors available on the market contain a tri-axial vibration sensor and a temperature sensor. One such sensor and two different mounting pads are shown in Fig. A.1.1. An installation with sensor mounted on an electric motor is illustrated in Fig. A.1.2. Tri-axial readings allow fault finding to be improved as comparing simultaneously collected data from the three different axes allows some failure modes to be more easily identified or discounted. Tri-axial sensors also reduce the overall cost per axis compared with single axis sensors without a major penalty in battery life, as simultaneous collection of the three axes requires the same power as a single axis (there will be a power cost in data transfer, however).

With the availability of Internet access to the data acquired by the wireless systems, anyone with appropriate access is able to view and analyze the data at any time and from anywhere, resulting in greater visibility and confidence in the faults detected and remedial actions proposed. Additionally, external experts are now routinely consulted as required for unusual readings which are not obviously caused by a single failure mode.

Using a number of wireless sensors to troubleshoot problem assets is also possible. The ability to fit them anywhere without the need for power or other cabling makes them a very attractive proposition if short-term intensive monitoring of an asset is required. Once analysis is complete the sensors can be redeployed to other problem assets as required.

Many systems are now starting to implement machine learning to try and aid the end user in fault diagnosis. This is currently at an early stage and should be used more as a guide than an authority at this point.

Fig. A.1.1: Modern wireless sensor with two mounting pad options.
Source: AESSEAL, Inc., Rotherham, U,K., and Rockford, TN.

Fig. A.1.2: A modern wireless sensor mounted on an electric motor.
Source: AESSEAL, Inc., Rotherham, U,K., and Rockford, TN.

A.1.6 Conclusion

Wireless sensors can be considered a useful addition to existing condition monitoring systems, but economically do not currently supplant the use of hand-held data collectors. The price of the sensors would need to reduce by at least a factor of 10 from current levels before their use on all assets becomes economically viable. Even then some labor cost would still be required to check the sensors are still fitted to the assets, replace broken sensors and flat batteries, and review the data.

Where wireless sensors excel is allowing the range of installed assets to be monitored. Wireless sensors can log to readings when no one is present; for example, due to batch processing or when machinery is inaccessible due to its location or guarding.

They can be used as a great troubleshooting tool where they can be quickly and cheaply deployed when a need arises. They can be recalled once the problem has been identified and solved, without the need for power and network cabling.

When choosing a wireless sensor the whole system should be assessed, not just capabilities of the sensor itself. The accessibility of the data over the Internet and the various features and management tools which may be included in the software system should also be used as selection criteria. Additionally, much weight should be given to the new abilities. To collect large numbers of readings requires improved capabilities to get the largest benefit from the installation.

A.1.6.1 The future of condition monitoring

Although it is almost certain that in the next two decades all machines will come with hardwired fixed sensors, there will still be a market for stand-alone condition monitoring systems. Highly efficient mobile data collectors used in conjunction with fixed wireless sensors where required will be around for some time to come.

The key ingredient will always be reliability professionals who can analyze the data and make and implement good reliability improvements. Think of software that collects data from a myriad of different sources and performs automatic analyses on asset condition, allowing reliability professionals to improve their plant performance and personal performance. Now *that* will be the future!

A.1.7 The sequel

While attending a pump-related technical conference and exhibition in 2018, the author was again made aware of the urgency of having our questions answered authoritatively. Our contributor, Dr. Ian Newby is to be highly commended to have done this in the above write-up. He understands that working with providers who see their task as merely selling sensors is quite inferior to working with a provider that sees the full picture.

At the 2018 conference, questions raised on the exhibit floor occasionally generated a response best characterized as mild annoyance with the ones who dared to "pour rain on the parade." It got us thinking that questions raised by a potential purchaser's reliability professionals would best be answered by respondents on the providers' management level. Hopefully, asking these questions will result in factual answers. Sensible answers must take into account the entire system and the qualifications of employees or contract personnel that use the system. Sensible answers would (mercifully) steer clear of the often unrealistic projections that reach far into the future.

Competent reliability professionals have always been strong advocates of PdM (predictive maintenance) and PdM will always have merit. However, for many decades moves have been underway toward placing progressively greater trust in predictive maintenance. But increased reliance on PdM is not very profitable in situations where failure avoidance by design would have been a much smarter strategy. Neither is merely relying on PdM profitable in hot and/or feed services and many other situations where equipment failures could create serious safety and environmental problems. Failure risks greatly escalate whenever several deviations from ideal component dimension, installation practice, and pump operating envelope combine, or when lapses in the quality of workmanship occur. While deviations of this type are infrequent, they nevertheless happen. The extent to which sensor technology addresses these concerns remains to be seen.

As just one example, trained professionals understand pump failure events. Slivers of brass indicate oil ring decay and oil rings are a component that works perhaps 95% of the time in pumps. When oil rings malfunction, they can cause rapid failures and for sensors to avoid these is unlikely. How to avoid this kind of contamination was explained in a 2020 text with the title "Optimizing Equipment Lubrication, Oil Mist Technology, and Storage Preservation," ISBN 978-1-941872-98-7

In conclusion, today's wireless sensor technologies have progressed well, and we hope that great successes will be reported as time goes on. Meanwhile, though, it would be prudent for reliability pros to work with technology providers that can explain how the issues that caused disappointments in earlier decades have now been circumvented or even solved. It's best to do our own digging and find out where a competent provider has installed a full-fledged system. There also is merit in determining if the reliability managers at such client locations have sought and obtained buy-in from their operators and field maintenance personnel. Once the fact-seeking reliability professionals are fully satisfied, they might ask highly experienced sensor technology providers to assist in compiling reasonably accurate cost justification calculations and case histories. Then it would be time for both satisfied user and competent turnkey provider to spread the word.

Appendix 2
Training reliability professionals and subject matter experts

As in all professions, there are individuals who graduated near the top of their class, and others who graduated last. Rarely would we want to entrust our health, home, children, or food preparation to the ones who "barely made it" and, since getting out from whatever schools they may have attended, have learned or contributed next to nothing. With that consideration in mind, we append this material to our main text for the benefit of managers whose responsibilities include the grooming or development of reliability professionals and/or subject matter experts (SMEs). We want these to be average, as a minimum, and above average, if possible. Also, this material is included to guide individuals who quite correctly view it as one of the roadmaps to self-improvement.

After a bit of history, this brief text segment presents experience-based remedies that have allowed many engineers to climb out of the rut in which others continue to find themselves. The specifics of successful training are given here. Specifics are important, because the engineering profession finds itself surrounded by buzzwords and clever acronyms. Many buzzwords and catchy phrases are mere consultant-conceived generalities. Some have been labeled "flavors of the month," the equivalent of feel-good quick fixes that add little or no value.

A.2.1 Tracing the origins of lack of recognition

In the 1950s and when some of us graduated from Engineering College in the early 1960s, a mechanical engineer's career was largely influenced by supervisors and managers who had moved through the same, or very similar, knowledge-based career steps. Guidance and direction given in those days was far more focused than what's offered five decades later by generalists and strictly managerial types. The world view of the early twenty-first century's boss is often shaped by motives, forces, and agendas that differ substantially from the ones which prevailed in 1960 or 1970. The ultimate effects of this "new thinking" are decisions that often endanger the health and safety of personnel. In some instances, decisions and indecisions have put the profitability and survivability of entire corporations at risk.

As we now enter the decade of the 2020s, far fewer engineers are enabled and empowered to be decision makers; also, the visibility of the profession has been diminished. A shrinking of the overall prestige of the engineering profession has taken place over the past 20 years. This reduced prestige is partly caused by engineers not objecting to a cheapening of products, and the profession often tolerating repeat failures. Add the fact that engineers rarely speak with one voice and have,

https://doi.org/10.1515/9783110674156-016

on more than one occasion, accepted sales pitches and claims that run counter to the laws of physics. As a result, even their very credibility is sometimes questioned.

Making matters worse, many engineers are either not interested in thoroughly documenting their work or are unable to express themselves with fluency and accuracy. These inabilities tend to impede their progress and can close the door to professional recognition. Then also, and quite unlike medical doctors whom one would consistently address as "doctor," engineers are perfectly content being called by their first names. That fact alone might illustrate why engineers are not held in the same esteem as the medical profession. Moreover, many engineering graduates consider studying for a Professional Engineer's exam a waste of time – for whatever reasons. However, having the drive and motivation to study for and pass such an exam would probably be perceived as a positive trait by others, and would often help improve one's own sense of self-worth and accomplishment.

In contacts with the legal profession, as expert witnesses or otherwise, engineers very often concede the entire field to the attorneys. Many engineers meekly endure attacks on their person or credibility and allow lawyers to browbeat them into submission. This, too, does nothing positive for the esteem in which others hold the engineers' profession. Allowing such attacks or conceding the field to those who shout the loudest, is not enhancing an engineer's image.

But, fortunately, it does not have to be that way. Observe, for instance, how the top companies train and grow their reliability engineers or subject matter experts (SMEs).

A.2.2 There are choices to be made

While time and unforeseen occurrences befall anyone, it is equally true that our lives are largely influenced by the choices we make. A reliability engineer can choose to receive virtually all his or her post-college training in the form of on-the-job learning. Of course, there's nothing wrong with traditional on-the-job learning; chances are the process reliability professional depicted in Fig. A.2.1 started his career by splitting his time between field training and the many other training methods that were available to him. We can be certain that one such "other method" will be chosen by those that buttress and supplement their field learning with mature reading habits.

We have seen Fig. A.2.2 on an SME's desk and can safely assume that he or she would endorse our observations. It is a well-recognized fact that mature reading habits will greatly accelerate the acquisition of thoroughly marketable skills. A disciplined reader's knowledge intake will progress in considerably more structured and measurable ways than the intake and absorption rate of a traditional on-the-job learner. Reading is one of the keys to professional growth. It would be unthinkable for a patient to prefer the non-reading medical doctor over her or his reading colleague.

Accepting and absorbing as "fact" what somebody else tells us may not always be of true benefit, but neither will the act of discarding everything that others have

Fig. A.2.1: An SME's training must include practical field exposure and considerable theoretical learnings.

Fig. A.2.2: Absorbing training material broadens experience; it can even prevent future disasters.

done before us. For instance, some engineers offer guesses instead of solutions, or accept on blind faith what others tell them "on the job." Some, not having trained their thinking abilities, confine their contributions to voicing concerns over potential problems. Although having a degree in engineering, practitioners of a shallow version of engineering put themselves at risk of being viewed by management as a dispensable job function, a "pair of hands," a person deserving little respect, one that can be readily terminated and easily replaced by contract employees.

On the opposite extreme of the ones who habitually spout off opinions are the perfectionists. Perfectionist engineers see doom in every decision made by man. They will "study things to death," not realizing that there are many endeavors that do not merit investigation or micro-definition beyond a certain point. Obviously, either of the two extremes must be avoided, and science must always trump gullibility and sales pitches. On the other hand, testing and understanding "the mechanics of things" is always a sensible pursuit, as is thoroughly examining underlying design principles. This implies that a balanced view must be sought; it stands to reason that finding and consistently practicing this balance will require effort. It also requires an investment in time, and certainly implies reading and thinking not just on one's employer's time, but also on one's own time.

Reading, learning, and progressing are shared responsibilities. Both parties must invest in them and the employer should periodically sit down with the employee to examine progress. During such performance appraisals, the parties could go over such topics as "how do we produce product 'X' or what procedure are we employing to get from 'A' to 'B' or repair 'Y' at this facility versus how the book tells us it's done at Best-in-Class ('BiC') facilities."

A.2.2.1 True professionals have balance

True professionals learn to identify root causes of problems and map out remedial actions that avoid problem recurrence. While it is certainly never too late to cultivate the balanced view, it is obviously best to do so early in one's life. Regardless of how and when acquired, a balanced view can be transferred into a job situation. The acronym "CCC" is worth remembering. It stands for Cooperation, Communication, Consideration. Practicing all three is needed if we want to consistently add value.

A.2.3 Why shared learning and a measure of specialization is important

When a person learns and/or adds experience in a field that is logically related to his or her job function, both employee and employer stand to benefit far beyond their original expectations. The employee gains a sense of self-worth that will allow them to confidently look ahead to an otherwise hazy employment future. By nurturing an employee's desire to learn and become an above-average contributor, an employer may well gain a value-adding employee. The employee adds value by acquiring and cultivating the ability to make "go-no-go" or "do versus don't do" decisions. These decisions will be based on more fully understanding risks and consequences; the ability to carefully weigh these decisions can be worth a fortune. A smart employer, therefore, makes training a shared responsibility. Such an employer will faithfully do his part; likewise, the employee will consistently and conscientiously do his or her part.

Bright people have an intuitive understanding of the merits of having not just a job; they wish to gain an increasing measure of marketable knowledge. They put themselves in charge of their own future and assign great value to the systematic acquisition of a definable specialty. They also know, ultimately, how they compare against above-average competitors.

A.2.3.1 Acting at the college student level

So, let's just assume the reader is a student and future mechanical engineer with the goal of specialization in rotating machinery for oil refineries and petrochemical plants, or *reliability improvement of fluid machinery* (pumps, turbines, compressors). Note that this arbitrarily chosen specialization goal is not as *narrow* as, say, "small metering pumps." An overly *narrow* area might not serve you in the long term if, for instance, small metering pumps were suddenly being replaced by "electronic stroking pistons" or whatever. Likewise, an overly *broad* area of specialization (such as "machinery and equipment") might be presumed to include bookbinding, consumer goods packaging, shoe manufacturing, and 10,000 other types of machines. Claiming coverage of such an area could be perceived as shallowness or naivete.

A.2.4 Specific steps in the training and learning processes

Just to re-emphasize: Accepting that the most important learning process *begins* at graduation is the first and perhaps most important realization in an engineer's training. While training plans will, of course, differ for different areas of specialization, it might surprise that the *principles* embodied in the specifics we chose to list here for "*reliability improvement of fluid machinery*" apply to every aspect of engineering specialization. The author's own experience supports the belief that meaningful training should progress in four or more distinct phases.

A.2.4.1 Phase 1: Reading trade journals

In the interest of continually obtaining work-specific technology updates and related training, the value-adding engineer must peruse trade journals. He or she should scan and retain (file) articles on topics of potential interest. We might use our imagination to interpret scanning as viewing information on paper or by way of an electronic scanner. One would then make copies of, or read, or tear out pages, from relevant trade journals. The material would be organized, filed and electronically catalogued for rapid retrieval.

Companies with well-defined training plans would involve all individuals who can benefit from the way these plans are implemented. The manager or supervisor would arrange for applicable *Trade Journal 1* to be given to employee “A.” In reviewing this trade journal, “A” notices an article dealing with shaft couplings and sends copies to colleagues/coworkers “B,” “C,” “D,” etc. Applicable *Trade Journal 2* starts on the desk of employee “B,” who notices articles on pivoted shoe bearings and wear-resistant V-belts. “B” makes copies of these two articles and sends them to “A,” “C,” “D,” etc.; likewise “C” sends articles of interest to “A,” ”B,” “D,” ” E,” and so on.

This once-per-month review task typically takes less than 10 minutes and allows each participant to acquire a data bank of relevant cross-references. The author recalls an experience when, decades ago, he looked for a reference article and then called its writer. He asked for---and cheerfully received---priceless guidance on a subject matter related to the writer’s topic. That kind of networking proved of great value and became the anchor point for other communications.

A.2.4.2 Phase 2: Technical books---a page per day, or 200 pages per year

Relatively few engineers purchase, or thoroughly read, technical texts after completing their formal education. Fortunately, however, there are some training-oriented employers who encourage their staff to read and absorb relevant technical texts. In 2003, one such employer explained that his responsible professional employees are encouraged to purchase as many books as they can reasonably assimilate or digest in a year’s time. During performance appraisals, the effectiveness of this policy was validated and reaffirmed on many occasions. Another company purchases pertinent technical texts and requires each technical employee to read a page per day. To the extent feasible and reasonable, these professionals are then asked to jot down what they discern as differences between *their* guidelines, work processes, hardware details, failure frequencies, maintenance intervals, etc., versus what *others* (and, in particular, competitors) are doing in these fields of endeavor.

The training value of this approach is immense, and is even more appropriate when used to add value while homebound. Reading and documenting can keep responsive employees involved; it allows them to grow towards becoming top-rated contributors upon returning to their plant. Certainly, the return on the investment of the time it takes to read a page-a-day and to make a two-sentence notation each week is huge. There can be no doubt that this well-focused training is priceless and benefits all parties for decades to come.

A.2.4.3 Phase 3, Part 1: Training through "Shirt-Sleeve Seminars"

In the 1970's and in pursuit of Phase 3 Training, one highly profitable multi-national company arranged for its equipment reliability technicians and engineers to share the responsibility of making seven to ten-minute presentations at the conclusion of each routinely scheduled and mandatory safety meeting. The presenters would educate themselves on such topics as "how to properly install a centrifugal pump," or "why steam turbines must be thoroughly pre-heated before full-speed operation." At the conclusion of his presentation, the presenter would distribute written copies of these one or two-page guidelines; they were handed out laminated in plastic and with one edge three-hole punched. Plant management made sure that reasonable expectations were met. Enlightened managers made it clear they wanted these guidelines to be read, consistently used, and fully adhered to. Reliability engineers, mechanical work force and operating personnel were told that compliance was mandatory, not optional.

In this manner, the "shirt sleeve seminar" presenters taught themselves by networking and communicating with knowledgeable people inside and outside the facility. As they made their presentation, they transferred their findings to the entire plant. At this location, equipment failures due to human error and other causes were minimized, and everyone profited from this approach. There should be no reason for not adopting it elsewhere with equal success.

A.2.4.4 Phase 3, Part 2: Presenting to the plant Steering Committee

The most successful process plants often have three Division Managers report to the Plant Manager (Fig. A.2.3). The Division Managers oversee and manage the disciplines Manufacturing, Human Relations, and Technical. Under the Manufacturing Division Manager, we have seen two department managers, one for operations and one for maintenance. These two managers are appointed with the understanding that they will be asked to switch their respective positions when asked to do so by the plant manager. This "hat-switching" may be ordered in random intervals of 3 to 15 months, with no more than eight hours advance notice. The effect will be that both managers are competent in each other's discipline. They will cooperate and never practice the destructive game of "one-upsmanship," the dangerous hoarding of information in "silos," and the unprofessional finger pointing that so often undermines trust and progress.

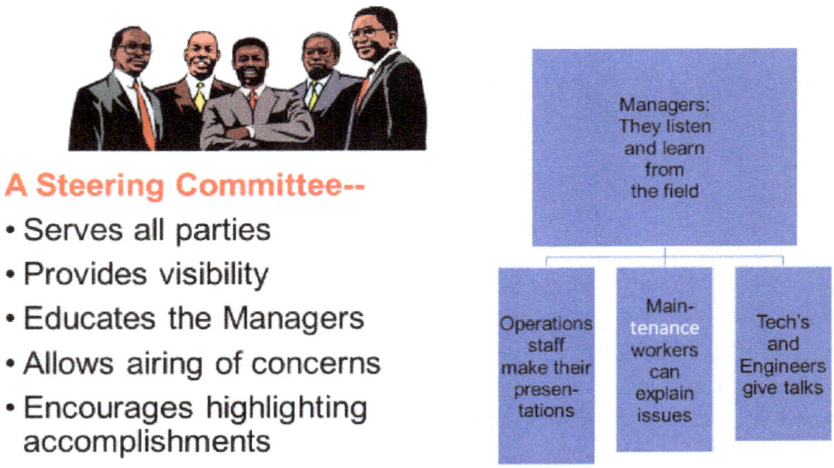

Fig. A.2.3: Functions of a Plant Steering Committee.

Reporting to the Technical Manager are Department Heads for New Projects, Reliability Engineering, and others. Section Heads report to Department Heads. Similar organizational arrangements report under the HR Division Manager.

The plant manager is supported by a Steering Committee whose members include the three Division Managers and several Department Heads. This Steering Committee invites staffers below the level of Section Head to make 10 to 15 minute presentations on how they do their job, what they would do different if it were within their power, what they would do if they were given $10 million with the mandate to spend the money for improvements, and how they personally have added value to the enterprise. These presentations monitor the pulse of an organization. They inform and even teach managers, enhance the sense of self-worth of the employee, and give high visibility to the presenter. Please don't lose sight of the end result of this phase, we called it Phase 3, Part 2.

A.2.4.5 Phase 4: Attending and participating in off-site meetings and Conferences

It may be noted that Phases 1 through 3 cost the employer next to nothing. The degree to which employees carry their load should determine the extent to which they will be involved in plans that include the next phase, Phase 4: Opportunities for offsite training. By mutual agreement between employer and employee, the employee may now progress to receiving this training. He or she would attend conferences and workshops that require fees, travel expenses, and accommodations paid for by the employer. After an employee returns from such offsite training, she or he will have to compose a 300-word report on things learned. They would have to explain how the learnings could be used at his or her place of employment. One facility found that it was best to combine the act of handing in this brief report together with the acceptance and timely processing of an employee's expense statement.

The employee's report is later disseminated throughout the company. Report writers benefit from refreshing their recollections, as will the various technical employee-recipients and their managers. Name recognition is established, and the contributor's sense of accomplishment is certainly enhanced in the process. Everyone benefits from the experience. The greatest hazard is that managers or supervisors are lax in their enforcement of having these one-page-maximum reports written, handed over, and issued. Non-enforcement sends the message that nobody really cares, and that the employer is happy with maintaining a below-average peer rating. It seems the employer's claim to optimizing the development of technical employees was deceptive and untrue. The employee feels betrayed and gullible because he fell for empty promises. Because he values those who value him, his next step may well be to find another job.

A.2.4.6 Role statements and future training plans

Already during a job interview engineers about to graduate from college would be wise to explore their projected role. Very soon after starting work, the new engineer should express strong interest in receiving a written role statement from his or her superior. Role statements define expectations. If no such statement is forthcoming, engineers are encouraged to put their understanding on paper and ask the responsible manager for review, input, recommended changes or concurrence. Unless there is agreement on the engineer's role, "performance exceeding expectation" is as improbable as the same person simultaneously dancing at two weddings located 7,000 miles apart.

Likewise, during the job interview, an engineer about to graduate should ask about the training opportunities made available or endorsed by the prospective employer. The interviewee must have a goal in mind and this goal must involve professional growth and learning. Learning is obviously a two-component process. While one party offers it and the other absorbs it, the ultimate benefits are shared by both. That being the case, each has a commitment to make and to honor. Serious forethought and cooperation are needed to achieve optimized professional training. And one more thing: Many opportunities will present themselves to the motivated technician whose formal schooling did not go beyond High School or past a two-year college. No worry, though, because a good non-degreed technical person will add far more value to an enterprise than an indifferent, often disruptive or highly opinionated graduate engineer.

A.2.4.7 Role statements are helpful to all

Again, role statements define expectations. As an example, a company could identify a self-motivated employee and ask this employee if he or she would be willing to be the custodian of an electronically stored and searchable engineering library. In the case of reliability professionals dealing with turbomachinery, pumps, gears, shaft couplings, etc., they would then be asked to identify useful Technical Conference Proceedings, published articles, and related information on the chosen range of topics. As mentioned above, the material needs to be indexed and, in one form or another, made accessible to one's peers and other individuals that would be helped by the reference material.

During performance appraisals, the employee and the manager, reviewer or performance appraiser would make an objective assessment of accomplishments by way of comparison with the previously agreed-upon role statement. Such an assessment would comprise all pertinent training issues and would obviously include measuring the employee's progress in terms of reading and disseminating technical material.

A.2.5 Favorable results anticipated

A.2.5.1 Acquiring a marketable skill

By accepting help and by being willing to assist others succeed, engineers will prosper. Moreover, they will gain an enhanced sense of self-worth if they truly pursue training that adds value. Engineers who succeed in acquiring a marketable skill both during formal studies and after graduating from engineering school can face the future with considerable confidence. Be sure to catch the two key words: marketable skill.

Self-motivated engineers who implement and stick to the approaches briefly described here are very likely becoming employees who offer solutions to problems. Instead of becoming folks expressing "concern" over potential problems, they will delineate the discrete steps needed to address these "concerns" by actions that will avoid problems.

Finally, then, here are a few more things to ponder:

- Certainly not all that is labeled *education* is beneficial. Some education can be so academic as to lack substance; it would not pass a balanced person's definition of a marketable skill. Training is a form of education. Take charge and make it relevant.
- Solving problems is more valuable than merely voicing concerns. Having a balanced approach allows us to see how far to go with an investigation, specification, or implementation.

- Don't just wait for skills-enhancing training opportunities to present themselves. Instead, take the lead in creating some of these opportunities. Virtually all marketable skills are acquired through proper response to meaningful training. True focus and the acquisition of *marketable skills* will get us through life better than a mere *education* or a smattering of unfocused, consultant-conceived flavors of the year.
- Selecting a profession, craft, trade or job simply because it is paying well is generally a mistake. Developing a passion for a job is equally inappropriate because spirituality and family should rank higher than a mere job.
- As a new employee and while on your way to work every day, make it your goal to *add value*. Think ahead, dwell on the specifics of *adding value* on that day. Then, on the way home from work, ask how successful you've been in adding value to the enterprise. If you have not met your mark or personal goal, resolve to make the needed adjustments on the following day.
- Remember that in your job you will encounter designated leaders that either cannot or will not lead. When this happens, don't give up. Only dead fish will always swim downstream. The flow is in only one direction, but a healthy fish will swim upstream as well.

Appendix 3
Specifying pumps for the oil and gas industries

Without a good specification, the purchaser sends the signal that price is the primary criterion in his selection process. In contrast, a reliability-focused buyer investigates life cycle cost or LCC. Both the author and the publisher believe that a "How to Do Pumps" appendix was desirable and so we enlisted the help of six EPC (Engineering/ Procurement/Construction) engineers, Paul Kenneth, Patowari Biswajit, Aymeric Housez, Rajarajan Ilangovan, Hervé Baron, and Ron Eierman, to compile this important text material. Their valuable core contributions and permission to liberally use or condense many of their up-to-date compilations are gratefully acknowledged.

A.3.1 Scope

The material in this Appendix 3 will equip non-specialists with most, if not all they need to know or understand about pumps in oil, gas, and – collectively called – hydrocarbon processing industry (HPI) services. The emphasis is on the key technical features and considerations when specifying a pump and checking that technical issues and realistic economic choices remain in balance. High asset reliability must be obtained throughout. Both centrifugal and positive displacement (PD) pumps will be discussed. PD pumps include reciprocating and rotary (screw, vane, lobe, gear) types. For each topic, criteria relating to both centrifugal and PD pumps are reviewed side by side. Although incomplete due to the vastness of pump subjects, this comparison will allow us to identify their relevant differences and similarities by resorting to [1] through [7].

Figure A.3.1 shows a centrifugal (overhung) pump. This is the type of pump most frequently used in the Hydrocarbon Processing Industry (HPI). It is safe to assume that close to 90% of the pumps in the HPI are of this type.

A.3.2 Inquiry documents issued for pumps

Inquiry documents are typically issued by an Engineering Company (often labelled EPC, which stands for Engineering Procurement and Construction provider). The EPC compiles and issues these inquiry packages on behalf of the owner-operator or ultimate purchaser. The bidder replies with a cost quote and detailed description of the equipment offered.

https://doi.org/10.1515/9783110674156-017

Fig. A.3.1: A typical back pull-out pump.

Within typically 30 days, the EPC issues to the selected bidder a set of pump purchase documents. These would become a binding specification of deliverables; for process pumps, the set of purchase documents usually consists of the material requisition and attachments, as follows:

- One data sheet for each pump
- A general specification, describing the Purchaser/Client's choice among the various options proposed in applicable American Petroleum Industry (API) standards, together with the Client's additions, substitutions, and/or modifications to API requirements, if any. On bulleted items listed in an API standard, the EPC must indicate which items to delete or to require. Not making this choice usually results in the vendor supplying least cost options
- An Inspection and Test Plan, or requirements related to inspections and tests (see dedicated section) and owner-purchaser's involvement
- A Required Documents list and schedule (which is also called "the deliverables")

The required contents of the Material Requisition (MR), which includes site atmospheric conditions (see later), power supply and utilities available, restricted subvendors list, etc., are contained in this MR.

It is useful to include in the MR a special notes section, highlighting the specific, stringent, and reliability-imparting requirements of the job. These requirements are indicated in the inquiry documents, e.g., on the data sheets, and reference should be made to these documents. However, it will be effective to group them in a dedicated section to emphasize their importance.

A.3.3 Choosing between centrifugal and reciprocating pumps

Centrifugal pumps should be selected wherever possible because they cost less and are more reliable. The only exceptions for not using a centrifugal pump but a positive displacement pump instead, are:

- very low flowrate, less than 3.5 m^3 per hour
- very precise (low) flowrate
- very high head, over 300 bar
- very viscous fluid, above 430 cSt, due to significant reduction in efficiency.
- Two-phase flow

Centrifugal pump speed (rpm) has traditionally been limited to 3,000 rpm (at 50 Hz) / 3,600 rpm (at 60 Hz) due to motor design. Higher speed pumps require a gear box or a Variable Frequency Drive. These high-speed centrifugal pumps can be an option when high energy must be imparted to the fluid, such as for high pressure water reinjection, etc.

The process engineer usually specifies the type of pump on the Process Data Sheet. How does he/she determine whether the flow + head combination is within the capability of a centrifugal pump? Figure A.3.2 depicts the head versus flow chart, and associated guidelines, to help the specifier fill out the process data sheet.

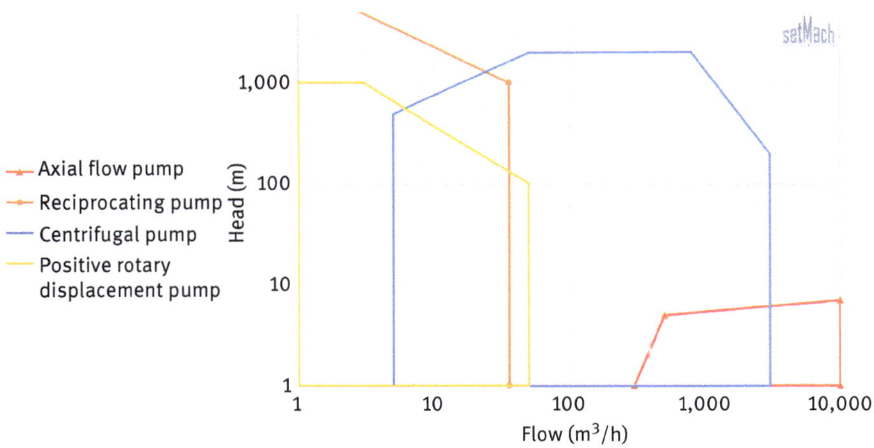

Fig. A.3.2: Pump Selection Chart.

A.3.4 Selection guide

As a general recommendation, the following are of interest:

1. Selection of pump type should conform to Fig. A.3.2, which shows the approximate upper limits of differential pressure and flowrate for the various classes of pumps. Rotary type pumps include screw, vane, gear, and lobe pumps.
2. Where a chart allows more than one class of pump, the selection sequence should favor centrifugal, rotary, then reciprocating, depending on factors shown below.
   a. Flow regulation
      Flow in centrifugal pumps varies widely with system resistance. In positive displacement pumps (rotary pumps and reciprocating pumps), the mean flow remains practically constant while the pumping energy depends on system resistance. If the service requires high flow stability, a positive displacement pump is the appropriate choice.
   b. Liquid viscosity
      Centrifugal pump performance deteriorates rapidly with increasing viscosity, and a PD pump is usually the better choice whenever the liquid viscosity exceeds 500 SSU (110 cSt).
   c. Energy consumption
      Displacement pumps are more efficient than the equivalent centrifugal pump, particularly those near the centrifugal pump upper limit, although with higher capital and usually higher maintenance costs. Consider the choice of PD pump when the energy cost is high.
3. Within the PD pump group, where both rotary and reciprocating pumps are allowed, the choice is rotary, then reciprocating, subject to two general limitations:
   a. Low viscosity
      Performance of rotary pumps falls off when liquid viscosity decreases because of an increase of internal leakage of the pump. If the pumped liquid has low viscosity, lower than approximately 100 SSU (22 cSt), or low lubricity and the differential pressure is high, a reciprocating pump is the more appropriate selection.
   b. Abrasives
      Rotary pumps have a low tolerance of abrasive solids in the pumped liquid because of their small internal clearances. Reciprocating pumps are preferred when liquids contain abrasive solids.

However, pump manufacturers have been extending continuously the range of centrifugal pumps so that it is best to consult them.

Positive displacement pumps include reciprocating pumps, which have pistons and plungers. There are also controlled volume pumps, which have adjustable piston stroke lengths, gear pumps of internal or outer tooth construction, vane, screw, and rotary progressive cavity types.

The most common types of reciprocating pumps have single or multiple pistons fitted with plungers in contact with the pumped fluid and moved by a crankshaft. Plungers are solid, not hollow, and are connected to the crankshaft outside the stuffing box by means of crossheads. The crankshaft is rotated at low speed by the electric motor through a reducing gear box. Multiple cylinders, e.g., duplex, triplex, are used to increase the flow. This also reduces the pressure pulsation because strokes are staggered. A duplex (multi-piston) reciprocating pump is shown in Fig. A.3.3.

A.3.4.1 API 674 positive displacement pumps – reciprocating

API 674 is the standard covering these pumps. They are used for moderate capacities and high differential pressures. They deliver a fixed flow; therefore, a Variable Speed Drive (VSD) must be used if the flow needs to be varied.

Fig. A.3.3: Duplex reciprocating pump.

“Metering” or “dosing” pumps are used when a small, precisely controlled volume of liquid must be delivered at moderate to very high pressure. A typical application is chemical injection, a service for which these pumps are well suited due to the wide range of materials from which they can be fabricated. Their design is inherently leakproof. The displaced volume is adjusted by adjusting the length of the piston stroke (Fig. A.3.4). The electric motor shaft has a worm periphery that rotates a gear in a plane that can be varied by a micrometer screw. The position of the plane determines the piston stroke length.

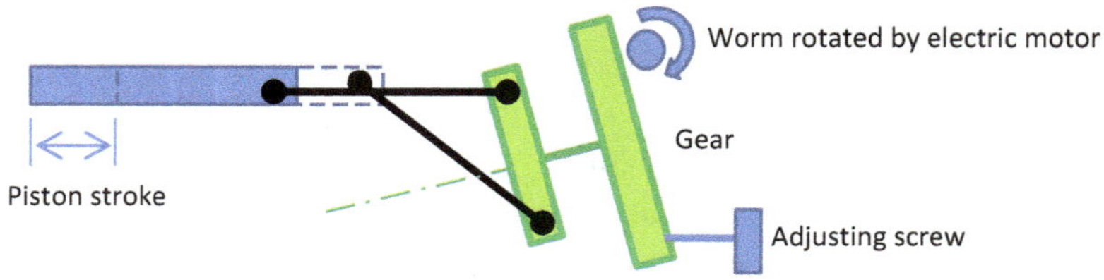

Fig. A.3.4: Principle of stroking mechanism in Metering Pumps.

The adjustment can be manual-mechanical or incorporating a fluid-filled actuator. A variable speed drive (VFD) can be used if flow variability is required.

There are two types of such pumps: The plunger type, wherein the pistons are in direct contact with the pumped fluid, and the diaphragm type, wherein the pistons are isolated from the pumped fluid. Diaphragm type pumps are used when the pumped fluid is of insufficient lubricity for the piston or when leak-tightness is required, e.g., for hazardous chemical or toxic liquids (e.g., ones containing H_2S).

API 675 positive displacement pumps – controlled volume is the standard covering these pumps. For further technical information about API 675 pumps, an interesting presentation can be found on the Internet: https://www.slideshare.net/vothanhtrung3/pump-api-675-presentation

Among the positive displacement pump, some use a screw, others use gears. They are generally used in highly viscous liquids or for very low flow for which centrifugal pumps and regenerative impeller pumps are not suitable. Low viscosity liquids with poor lubricating properties (such as water) are not a suitable application for gear or screw pumps. However, regenerative pumps (see Internet for details) and canned motor pumps are often capable of handling such fluids.

A.3.5 Centrifugal pumps and highlighting their main parts

The main parts of a back pull-out style process pump are identified in Fig. A.3.5.

Not shown is the coupling hub; it is fitted to the keyway on the shaft end.

A.3.5.1 Sparing options

Installed spare or pull-out assembly in store house require study and communication. To better understand sparing options one should ask, and answer, two important questions:

1. Why spend capital funds for installed spare pumps and why deal with issues associated with operating pumps in parallel are the most frequently asked questions.
2. Is an installed spare pump required for process safety or to ward-off significant production losses?

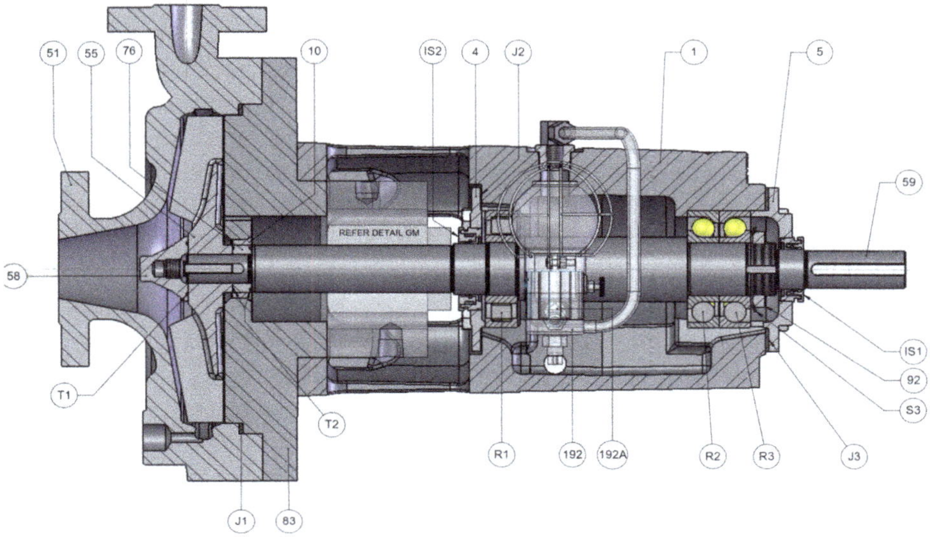

Fig. A.3.5: Cross-section of a back pull-out pump with open-style impeller.
T1 = Open style impeller; T2 = Spacer; 1 = Bearing housing; 4 = Inboard bearing end cap; 5 = Outboard bearing housing end cap; 51 = Inlet nozzle; 58 = Impeller nut; 59 = Keyed shaft; 83 = Back cover; 192 = Constant level lubricator with balance line; R1 = Roller bearing; R2 = Back-to-back thrust bearing set; J3 = Flat gasket; S3 = Tab washer; 92 = castellated nut.

EPCs rarely ask these questions in Pre-FEED, FEED (Front End Engineering and Design), or HAZOP. Owners and their design teams tend to miss optimization opportunities by not addressing the two questions in a timely manner. Sometimes the answer is self-evident but at times it requires research. The findings should be discussed with all the stakeholders.

There are many refineries and chemical plants where NO SPARE PUMPS were installed UNLESS process safety or profitability were significantly impaired. On average in these process units only 10% of the pumping systems were provided with installed spares. With some notable but unspecified exceptions, all pumps were centrifugal impeller canned motor pumps (seal-less, canned motor, see Chapter 10) or magnetically-driven positive displacement pumps.

The CAPEX (Capital Expenditure) total-installed cost savings on pumping systems exceeded 75%. Today, in 2020, chemical processing units so equipped have the highest reliability and availability in the specialty chemical sector at the Operating Company (one of the world's largest), and the fewest forced outages with no fugitive emissions from pump seals. Remember that there are no mechanical seals in canned motor pumps.

A.3.6 Centrifugal pump type selection

Pump vendors select impeller diameters suited for the design conditions but a pump body can usually accommodate a larger impeller. It might be useful at some point in the life of a pump to change the impeller to a larger one, for instance if process conditions change. Hence the client's (owner's) and EPC's customary recommendation is to take an available maximum diameter impeller into account for sizing the motor.

In case there is no client requirement, leave the choice of the pump type selection to the vendor. Ask for competitive pricing and yet insist on reliability. Find the true optimum and never compromise safety.

The smallest, and often claimed both least expensive and most efficient, is the vertical inline overhung pump, OH3, Fig. A.3.6. The claims, relating to initial cost and power efficiency, are not always correct. The same or better cost-reliability-efficiency-availability can be combined into life cycle cost. It then has often been shown optimum in canned motor pumps produced by the most experienced high-quality canned motor pump manufacturer(s).

Fig. A.3.6: Vertical inline overhung pump, OH3.

An in-line pump (Fig. A.3.6) is a pump designed to be installed in-line, not unlike a valve. It saves footprint, simplifies piping, and occasionally requires no foundation at all. The motor to pump connections can be rigid (close-coupled), which decreases reliability compared to horizontal overhung pumps shown in Fig. A.3.5. In the Fig. A.3.6 design, the radial bearings are usually located in the motor, which results in higher moment and wear for the bearings and mechanical seals. Hence, they require more frequent servicing.

The primary use of close-coupled and rigid-coupled in-line pumps is for light duty, non-critical processes or water service with low head and low flow. Their power is limited to 200 HP at 3,600 rpm due to potential vibration concerns and to 200°C (392 deg. F) pumped fluid temperature due to prevailing shaft sealing, bearing cooling and motor cooling.

A notable exception is for oil mist-lubricated bearings. Also, there are vertical pumps with longer (7-inch) distances between motor and pump shaft ends. If properly installed and with suitable vibratory characteristics designed-in, these superior verticals have been used with drivers up to 1,000 kW and reliabilities identical to those of horizontal pumps.

The relative advantages and disadvantages of vertical and horizontal pumps are shown Table A.3.1:

Tab. A.3.1: Comparison of popular pump styles.

Advantage (+) / Disadvantage (−)	**NPSHr**	**Footprint**	**Headroom**	**Maintenance**
Horizontal	More (−)	More (−)	**Less (+)**	**Easy (+)**
Vertical	**Less (+)**	**Less (+)**	More (−)	Less easy (−)

The next higher cost pump is the horizontal overhung OH2 shown in Fig. A.3.5 and Fig. A.3.7 It is also the most commonly used type or style of pump.

Sometimes this selection is not possible due to non-availability of an OH of the required size (discharge and suction diameter, impeller diameter) or appropriate speed. Limitations of OH pumps are often related to vibration; this can be caused by tip speed, hence by the product of impeller diameter times shaft rpm. In low pressure, low temperature and low power applications, BB4 type (between bearing-style) pumps, made up of identical stages that are "sandwiched" together, Fig. A.3.8 could be selected.

BB3 is the next cost-effective choice. It costs 20% to 25% more than BB4. It is horizontally (axially) split and therefore easy to maintain as the top cover of the pump can be removed by lifting. BB3 can be used up to 100 bar and 200°C only.

A vertically (radially) split casing BB5 type (Fig. A.3.9) is required beyond the above pressure and temperature limits. They are easier to maintain because their

Fig. A.3.7: API-compliant horizontal overhung (OH2) pump.

Fig. A.3.8: Horizontal between bearing (BB4) pump.

cartridge design allows quick replacement. However, BB5 types are 15% to 20% more expensive than BB3 types.

High speed pumps, described in API 610 as OH6, are manufactured by a few specialized vendors. Vertically suspended or "VS" pumps are top mounted and have their liquid ends submerged in the pumped liquid. These are the VS1 to VS7 according to API 610 which are primarily used for service to improve NPSH available (see NPSH) or for service whose suction pressure is close to the liquid's vapor pressure. These pumps offer the advantage of being able to stack multiple impeller so they can reach very high heads. VS pump installations cost less if mounted on their suction vessels.

Figure A.3.10 depicts the most likely (or generalized) selection envelopes for defined pump types; however, there are no strict limitations.

Fig. A.3.9: BB5 Pump, between bearings, radially split, centerline-mounted.

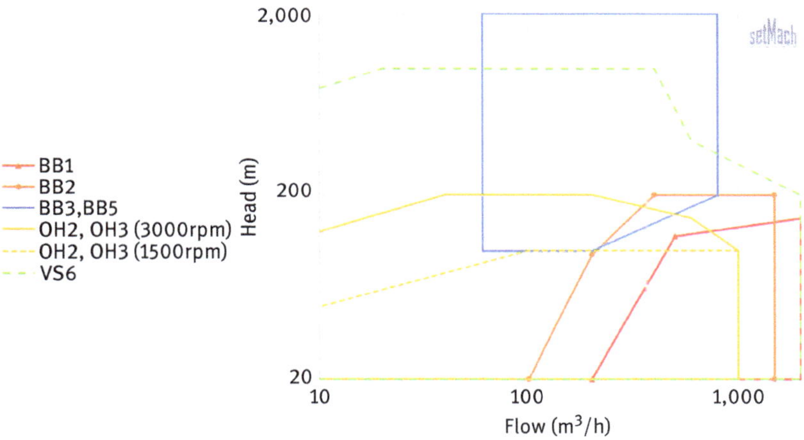

Fig. A.3.10: Generalized selection envelope for typical centrifugal pumps.

Self-priming centrifugal pumps are used in a suction lift arrangement; in a lift arrangement the liquid to be pumped is at lower elevation than the pump's suction nozzle. Self-priming pumps have an enlarged casing, allowing the end suction nozzle to be placed in an offset position above the impeller. The impeller is thus always kept submerged and creates the necessary vacuum to draw the liquid in. One would specify self-priming pumps when a pump cannot be located below or on top (VS type) of the suction vessel. This pump is normally used in a liquid with ample suction pressure or, more precisely, where considerable NPSHa (net positive suction head) is available. Its application is mostly in sump pits and for the purpose of transferring water to other storage facilities. Horizontal direct drive self-priming pumps should be used for open sump services when suction lift (vertical distance between the pump's centerline and low-low liquid level in the sump) does not exceed 6 m (20 ft).

There are two caveats for self-priming pumps:

1. A self-priming pump requires initial flooding of the casing before start-up. This initial flooding is achieved through a normally plugged threaded hole on the casing.
2. The discharge piping should be equipped with an air vent or air release line. This line is required to remove air from the casing during initial start-up.

The most common type of self-priming pump has the requisite cavities cast in the casing. However, there also exist self-priming pump designs without the casing cavities. These pumps achieve self-priming through a combination of suction piping design and special casing design. (For details, consider using the link below: http://www.aquadevice.com/english/03pump_uhn7.htm)

A.3.7 Process specification

Either a stand-alone process data sheet is issued by the process designers, or these designers fill-in the process information sections of the API data sheet. Among the information to be provided by process designers, the following should be included:

- NPSH

  The purchaser specifies the available NPSH ("NPSHa"). This is the margin, expressed in feet or meters, that exist between the pressure of the liquid at the pump suction, and the pressure at which the liquid would begin to vaporize. The NPSHa is calculated by the process engineer, taking into account the elevation of the suction vessel, its internal pressure, the nature of the pumped fluid, and the friction losses in the suction pipe.

  It is important to perform this calculation task precisely. It is imperative to guard against an uninformed person adding what they consider a safety margin. Specifying too low a NPSHa could lead to pump oversizing by vendors, usually done by increasing suction nozzle inlet size, often done without understanding the unintended consequences that manifest themselves as destructive internal recirculation.

  It is recommended to specify a margin of a least 1 meter between NPSHa and NPSHr at the BEP (Best Efficiency Point). For pumps with auto-start or in parallel operation, specify additionally that NPSHa-NPSHr margin at end of curve should be 0.5 meter minimum. However, these guidelines are not applicable to certain aqueous services (think ammonia) where differences between

NPSHa and NPSHr as high as 20 ft (6 m) are sometimes required. Inadequate NPSH margin can cause serious cavitation damage. Be sure to investigate!

A.3.7.1 Pumps operating in parallel or with Auto Start or open discharge valves

There can be compelling reasons why pumps operating in parallel should be avoided. If one must use pumps in parallel, be thoroughly aware of additional constraints. These would include:

a. Pumps must have identical head/flow characteristics so as not to drive each other into an unstable operating range. They must typically have a head rise from rated to shutoff of at least 10%.
b. For centrifugal pumps in case of parallel operation, for example, pump A/B/C – Pump A, B in parallel operation and C is standby. In case of one pump failing, the standby pump will take some seconds to be back in operation, and the motor sizing of the one pump operation shall be consistent with total system resistance curve (provided by the process designer).
c. Pumps operating in parallel must be able to operate at "end-of-curve," which requires additional motor power (see later).

A.3.7.2 Hazardous area classification

As the process design engineer knows, from the properties of the pumped fluid it is possible to define the needed hazardous area classification. This classification derives from the composition of the pumped liquid (transferred into vapor) and from the ignition temperature of the various hydrocarbon fractions.

The area classification is defined by possible sources of leakage during normal operation (generally the mechanical seal regions). Hazard area coverage is usually a function of shaft size and speed.

a. Suction and discharge pressure:

   The operating suction and discharge pressures are indicated on the Process Data Sheet. The discharge pressure is estimated at the time when the pump is specified and ordered. Because the exact elevation of the downstream equipment and piping length/routing are not yet firmed up, a check will have to be done later to confirm these hydraulics.

   What also needs to be carefully specified are the *design* suction and discharge pressures. The design suction pressure is the design pressure of the suction vessel + the hydrostatic pressure between the suction vessel and the pump, i.e., height between vessel normal liquid level and pump centerline in meters, all multiplied by specific gravity.

The design pressure of the discharge piping must be specified; this will enable the vendor to propose a suitable pump. The discharge pressure of the pump, which at zero flow is called shut-off pressure, must not exceed the design pressure of the piping. Otherwise either an increase of the discharge piping design pressure or a PSV would be required, resulting in extra cost. Other industries limit the head rise from rated to shutoff pressure to 20% maximum. A pressure 20% higher than the pump's rated pressure is used as the maximum allowable working pressure of the downstream equipment.

b. Fluid conditions:
Maximum viscosity and density, usually at the lowest anticipated ambient temperature, should be specified by the purchaser. The pump and its driver must be designed and selected to start under these most severe anticipated conditions.

c. Capacity
Process designers indicate both the operating flowrate and the design flowrate. The difference, called "overdesign," is normally 10%, except for intermittent or recirculation service for which it is 0%, and for reflux and pump-around services where it is 20%.

The following concerns centrifugal pumps only:

d. Vapor pressure:
The pumped fluid *vapor pressure* should be specified. This will enable the vendor to ascertain that the fluid can be used for sealing. Its vapor pressure must be such that even with added heat input from seal friction there will be no vaporization. A dual seal and suitable barrier fluid environment must be selected to eliminate the risk of vaporization in the seal faces.

A.3.8 Capacity control

Pump performance curves are shown in every pump text, among them [3] and [4] They show that the flow through a reciprocating pump cannot be varied, as indicated in Fig. A.3.11, left side. The greatly exaggerated trend curve on the right side of Fig. A.3.11 depicts the head versus flow relationship found in centrifugal pumps.

The capacity (flow) of a centrifugal pump can be easily adjusted by a control valve in the delivery circuit. This control valve changes the pressure drop of the circuit, resulting in the change of flow through the pump (the pump pressure and throughput volume operate along the performance curve submitted by the vendor). To allow proper control, specify "head at shut-off to be at least 110%, but not more than 120% of rated head."

The flow through a reciprocating pump cannot be varied in such a way and, if flow variation is required, on-off control, recirculation (under flow or pressure control) to the pump suction vessel or a Variable Speed Drive (VSD) varying the electric power frequency, is required. If recycling is selected, the Tab. A.3.2 compares the types of capacity control available for each pump type. The selection should be based on the requirements of the process.

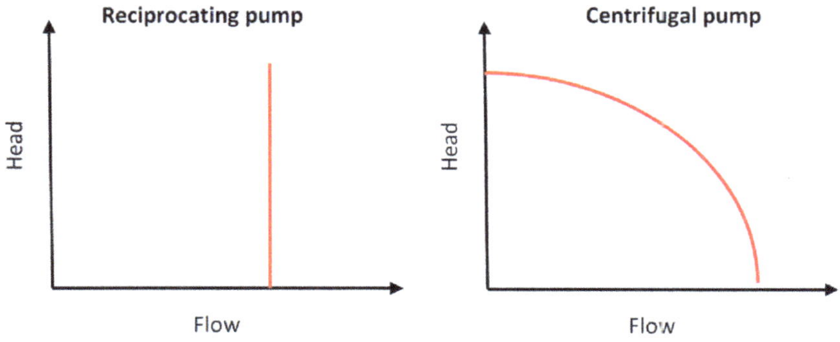

Fig. A.3.11: Flow versus head trends in positive displacement pumps (left) and centrifugal pumps (right).

Tab. A.3.2: Different capacity controls in use.

Type of pumps	Throttlingcontrol	Speedcontrol	Strokecontrol	Bypasscontrol	On-offcontrol
Centrifugal pumps	✓	✓	–	✓	✓
Reciprocating pumps	–	✓	✓	✓	✓
Rotary pumps	–	✓	–	✓	✓

Note that for a centrifugal pump, a minimum flow line with a control valve can be used for both functions of ensuring minimum flow and throttling.

A.3.9 Electric motor drivers

Normally, electric motor ratings relate to ambient temperatures of 40 °C and site elevations of 1,000 meters. Deviations from these conditions may require derating of electric motors.

A.3.9.1 Starting an electric motor

The easiest and most common way is just to apply the full line voltage to the motor. This is called DOL (Direct-On-Line). DOL starts cause a high starting current to flow.

Eight times the nominal current is rather typical, and the power supply system may not be able to provide it. Moreover, high currents can stress the driven application, and when this is to be avoided, at least two options present themselves:

1. Use a motor with star-delta (Y/Δ) wiring. This reduces the starting current by 30% compared to DOL. It requires two different power cables to the motor, they are connected to different terminals. This option is used infrequently.
2. Use a soft start feature which gradually increases the voltage.

Motors for DOL and star-delta start should be rated to carry the load and be able to start at 80% voltage. This is a standard practice and plant electrical supplies can operate within this range.

Note that motors with variable speed drive commonly use a frequency converter device. These devices do not have start-up current issues; they make "soft" low-current/low speed starts possible.

After purchase order (PO) award, the motor vendor should clearly state, on the motor data sheet, the starting time required. The vendor should provide these data based on the load information provided by the pump manufacturer or vendor. A motor toque/speed curve with superimposed load curve is required.

A.3.9.2 Motor power factor

The power factor, cos φ, i.e., the ratio between active power actually converted into mechanical work and reactive power used for magnetization, is usually between 0.7 and 0.9. There is little difference between motors used in the HPI and motors used in other industries. As the thinking goes: Where energy is cheap, there is little incentive to pay attention to this factor.

Axial pumps (also called free-standing blade or propeller pumps) must be started with the discharge valve open, i.e., to be able to operate at end of curve. This is where this pump style has its lowest power demand.

Contrary to a reciprocating pump, a conventional impeller-style centrifugal pump can be started with its discharge valve closed. This is the best way to start the pump because it drains less power from the driver. However, to allow easy opening of the discharge valve, it is customary to open the valve off its seat, also called "crack open."

A centrifugal pump shall only be specified to be able to start with its discharge valve open if the pump is part of an auto-start logic, e.g., spare pump P-1B automatically starts if P-1A fails.

A pump in auto-start would run to its end of curve when started because the discharge valve is fully open. End of curve operation requires more power from the motor; the motor must be sized accordingly.

Do not specify the ability to "start with discharge valve open" or "operation at end-of-curve" which has the same meaning, unless the pump is in auto-start logic as it would imply oversizing the pump driver. Note again that axial "propeller" low head/high flow pumps draw more power at zero flow than at end-of-curve (maximum) flow. Accordingly, the must be started with an open discharge valve.

A.3.9.3 Best practices for sizing pump drivers

The authors' company recommends obtaining an early agreement with the Operating Company on a few guiding principles. These include, during the pre-FEED stage:

1. Determine the maximum diameter impeller for the selected pump
2. Determine the maximum specific gravity of the pumped fluid. This may be water during commissioning,
3. Select the driver sizing that will be acceptable for items (1) and (2) above.
4. Discuss this sizing with the Project Team as they will be concerned about incremental CAPEX,
5. Discuss the driver sizing with Operating and Maintenance Teams as they desire a reliable asset and are concerned with OPEX and lost production,
6. Present the analysis and data to an Operating Company Management Team for final decision,
7. Ascertain that the amounts budgeted for machinery are assuming the procurement is for reliable machines, not lowest cost machines.

Once these guiding principles are defined, they are easy to follow through.

A.3.10 Applicable codes

The selection of the design code is an important factor determining equipment cost. The major differences between the two codes are listed in Fig. A.3.12. As an example, centrifugal pumps equipped with mechanical seals, API 610, the standard for centrifugal pumps in the Petroleum, Petrochemical, and Natural Gas Industries, may not be required for all services. This standard indeed includes numerous requirements such as casings designed for 40 bar, mechanical seals, etc. API 610 specifies its scope of cost effective application in its foreword: High pressure (discharge pressure > 19 barg or suction pressure > 5 barg) or high temperature (pumping temperature > 150°C) or head > 120 m. Outside these criteria another standard, ASME B73 Centrifugal Pumps for Chemical Processes, could be selected, even for hydrocarbon services, if sufficient experience exists in equivalent services and if the desired reliability has been ascertained. However, reliable and low maintenance services will require use of mechanical seals. Although mechanical seals are not a requirement of ASME B73, they must be provided in the interest of low maintenance.

	ANSI/ISO (ASME B73.1/2, ISO 5199)	API (API 610 = ISO 13709)
Standard dimensions	YES	NO
Casing pressure rating	20 bars	40 bars
Service conditions		Exceed one of the following: •Discharge pressure > 19 barg •Suction pressure > 5 barg •Fluid temperature > 150°C •Rotative speed > 3600 •Rated total head > 120m •Impeller diameter (OH pumps > 330mm
Mounting feet	Bottom	Center lined

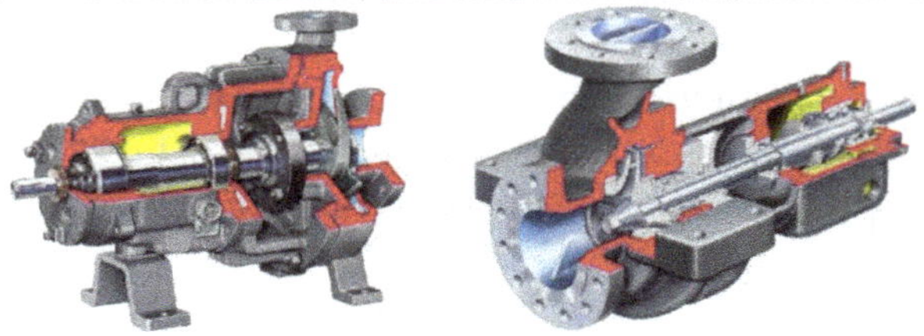

Fig. A.3.12: Comparison of the two primary pump standards used by process industries.

One advantage of ASME B73 is that it imposes standard dimensions to the pump so that it is interchangeable, regardless of who supplied these pumps.

API 685 is the standard applicable to seal-less pumps, which are of two types: magnetic drive and canned motor drive.

The electric motor and impeller of a canned pump, shown in Fig. A.3.13 in light blue, are on the same shaft without coupling. The pumped liquid circulates in the motor to ensure lubrication and cooling (see Chapter 10).

In a magnetic drive pump, shown painted in dark blue in Fig. A.3.13, the impeller is magnetically coupled to the electric motor. The electric motor rotates an outer magnet which faces an inner concentric magnet on the impeller shaft. As the outer magnet rotates, so do the inner magnet and the impeller.

From a Process point of view the type does not matter, the choice can be left to the Vendor. For less severe service conditions, **ASME B73.3** "Specification for Sealless Horizontal End Suction Metallic Centrifugal Pumps for Chemical Process" or manufacturer standard should be specified.

The three API standards applicable to positive displacement pumps are

1. API 674 Positive displacement pumps – reciprocating
2. API 675 Positive displacement pumps – controlled volume
3. API 676 Positive displacement pumps – rotary

Fig. A.3.13: Canned motor pump (lower image) and magnetic drive pump (upper image)

Aside from the selection of the applicable API standards, certain additional standards or codes may be specified for auxiliary systems. For instance, pump lube oil system, sealing system, gears, and couplings could be specified to comply with the relevant API standard rather than a manufacturer's standard. Please refer to dedicated sections. Major components of the auxiliary systems may themselves require compliance with API standards; manufacturers' standards often favor price over long-term quality.

A.3.10.1 Use of magnetic drive pumps

Seal-less centrifugals are cost effective where tight sealing is required, which implies the use of dual mechanical seal for a conventional centrifugal pump. This is particularly true for small pumps, typically below 10 kW, where the cost of the double seal support system occasionally becomes higher than the cost of the pump itself.

Some users restrict the use of magnetic drive pumps to services with pump power inputs below 55 kW, although many end users have updated their specifications based on proven experience from vendors. The actual limitation of magnetically driven pumps greatly exceeds 55 kW. Of course, experience with canned motor pumps now reaches up to 750 kW (see Chapter 10). In any event, seal-less pumps are worth studying. An additional benefit of both magnetic drive and canned pumps is the absence of fugitive emissions.

A.3.10.2 Type of factory assembly

Pumps are best delivered mounted together with the drive train on a common baseplate/skid. This is specified in API 610 and API 685 but left open in API 674 and other standards. It should therefore be specified by the Purchaser as "Complete unit mounted on single skid/baseplate." However, pump and driver must be removed during the baseplate installation process on a concrete foundation. The flatness of the baseplate mounting pads can only be verified by having full access to these pads.

If provided, the pump's seal system or sealing auxiliaries should also be installed on the pump baseplate or skid. The type of baseplate (grouted versus non-grouted, also available epoxy-filled) must be specified. The baseplate will not be grouted if the pump is installed on a steel structure but will be grouted if the pump is installed on a concrete foundation. Non-grouted installations require stiffer baseplates, hence are more expensive. However, non-grouted but pre-filled with epoxy is often lowest cost, acts as a stiff monolithic block, and is by far the least maintenance intensive. Become familiar with this option by consulting the references.

Pumps with long motor-plus-pump assembly or high energy used in offshore applications are specified with three-point support so that the pump and driver always remain in a geometric plane. Indeed, offshore structures are subject to bend due to sea motion. This increases the cost because the baseplate must be stiffened. Unless shafts are always horizontal, oil ring lubrication becomes risky. Recall that solid discs are often far less prone to fail than oil rings [2] and [7].

A.3.10.3 Liquid collection

When leakage from the pumped liquid to the foundation is not acceptable, e.g., for fire, corrosion, or pollution reasons, leakage must be collected. This is done using either a drain rim or a drain pan. A drain pan is a plate welded inside the baseplate that slopes to a drain point. This ensures that any leaks drain without accumulating underneath the leakage point. A drain rim is a less expensive design; leakage overflows travels outside the baseplate to channels welded to the edge of the baseplate. Due to the absence of slope, some leakage accumulates underneath the leakage point.

A.3.10.4 Material of construction

API 610 does not mandate the materials of construction. It only gives general guidance for usual services in one of the appendixes to API 610.

The material of construction of each part of the pump is further defined in an API 610 appendix for each class.

A.3.10.5 Metallurgy for sour service

All materials of components exposed to wet hydrogen sulfide in concentrations exceeding the limits prescribed by NACE MR0175 (Upstream) or NACE MR0103 (Downstream) should incorporate the reduced hardness requirements of NACE MR0175 or NACE MR0103. However, this stipulation does not apply to stainless steel grade, since the required hardness (<RC 22) is already complied with.

A.3.10.6 NPSH

To re-emphasize, the owner-purchaser specifies the available NPSH. It must be done precisely because too low an NPSH could lead to the pump being oversized by vendors. Oversizing inevitably implies a large suction to reduce inlet losses. Customary practice is to specify a margin of a least 1 meter between NPSHa and NPSHr at rated point. For a pump with auto-start or in parallel operation, specify additionally that NPSHa–NPSHr margin at end of curve shall be 0.5 meter minimum. But again: These values are totally inadequate for aqueous fluids, especially ammonia.

A.3.10.7 Transmission

Centrifugal pumps do not need a gearbox because the pump is designed to rotate at the same rpm as the electric motor. The connection between the shafts of the driver and pump is made by a simple non-lubricated, flexible disc coupling.

Reciprocating pumps rotate at low rpm, usually much lower than that of the electric motor driver. A speed reducer is usually required between the electric motor shaft and the pump shaft. Speed reduction can be achieved, for low power pump applications, by means of gears integral to the motor drivers (allowed by API 674 up to 18 kW). For low power, it can be achieved by means of a belt coupling the driver and pump shafts (allowed by API 674 up to 150 kW). However, belt drives require maintenance; they are rarely a wise choice in modern plants. For larger machines, a gearbox is required; it would normally be connected to the pump shaft by means of a low speed coupling and to the electric motor by means of a high-speed coupling. API 674 specifies that the gear box shall be compliant with the relevant API standard (API 677 or API 613).

The requirements for the coupling are specified in the API standard of the pump itself, i.e., API 610, etc. Reference to another standard is not necessary.

Tab. A.3.3 summarizes the transmission types for reciprocating versus centrifugal pumps.

Tab. A.3.3: Comparison of Transmission Arrangements.

Type of Pump	rpm	Coupling between Driver and Pump	Specifics of Drive Arrangement
Reciprocating	low	Reducer	integral gear, belt up to 150 kW, gear box above 150 kW
Centrifugal	high	Direct	simple non-lubricated couplings; however, API-style pumps require disc pack couplings

A.3.10.8 Sealing

The space between the rotating shaft (in a centrifugal pump) or the piston (in a reciprocating pump) and the pump casing must be sealed. The purpose of the seal is to retain the pumped liquid inside the pump at the point where the drive shaft or piston penetrates the pump body.

- Reciprocating:
  For reciprocating pump, sealing is quite easy and is done using packing. Double packing is specified with recovery of leakage and recycling to the suction of the pump for hazardous/toxic fluids.
- Centrifugal:
  For centrifugal pumps, sealing is more complicated than for reciprocating pumps. Gland packing, as used for sealing valve stems, may only be used for ASME pumps in non-hydrocarbon service. Even then, such packing imposes a maintenance requirement that is out of tune with the reliability and safety thinking of modern facilities.

A mechanical seal should be specified for hydrocarbon services and is required by API 610. Mechanical seals are purchased from specialized suppliers and fitted to the pump. As per API 610 they must be cartridge type, meaning completely self-contained.

Although many flush plans route a small side stream from the pump discharge through the seal and back to pump suction, the cartridge is occasionally fed by fluids from a rather complex external system, which can take up a significant amount of space. Recall that Chapter 6 of this text gives excellent guidance on how to always obtain the best possible mechanical seal.

The sealing fluid may require cooling through added equipment: Finned pipe, possibly a Forced Draft Heat Exchanger or a shell and tube exchanger using cooling water have been used in industry for many decades.

Generally speaking, a mechanical seal consists of two rings, a stationary ring fixed to the pump body, and a rotating ring fixed to the shaft. The rings face each other and are separated by a thin liquid film. The liquid cools and lubricates the seal faces. It must be selected so that it does not vaporize. Vaporization would damage the seal faces.

In a single seal, the liquid used is usually the pumped liquid, taken from the pump discharge (Flush Plan 11). If subject to vaporization, which must not be allowed to happen in the seal, the seal flush liquid is routed through a cooler by a pumping ring prior to being introduced to the seal faces (Flush Plan 23). Consult Chapter 6 for more detail and pay close attention to superior technology offered by competent seal manufacturers. Emphasize service and quality, not cost.

If the process fluid is not suitable for lubrication, for instance if it could contain solids, an external fluid is used (Flush Plan 32). It must be compatible with the pumped fluid because a small fraction will leak into it. Whenever possible, water is typically used as an external flush liquid. Water will remove more heat than oils and other hydrocarbon liquids.

In a single seal arrangement, a small quantity of the sealing fluid leaks to atmosphere. Usually the fluid used is sufficiently volatile so that the heat resulting from fluid traversing the seal face causes the leakage to vaporize. Such fluids may display no visible leakage. Other liquids will not so readily vaporize and thus may show an accumulation of leaked liquid outside the pump. In this case the typical leakage rate is approximately 10 drops an hour.

When leakage of the pumped liquid to atmosphere is not acceptable due to the vapor being hazardous (i.e., flammable or toxic), a secondary seal is added to the primary seal described above. The leakage of the pumped fluid through the primary seal is collected in the space between the two seals and its vapors are routed to the flare. The secondary seal uses a buffer liquid stored in a seal pot. The fluid is circulated by a plain pumping ring or helical screw-like auger. There also are innovative bi-directional tapered rings (Chapter 6).

Cooling can be added, if required, by means of a cooling coil inside the external sealing pot. The level of buffer liquid in the external pot is monitored and the pot refilled when required. This type of arrangement is called dual unpressurized. The space between the two seals exists at a pressure lower than that of the seal chamber.

With dual unpressurized seals, some pumped liquid, if *not volatile enough* to vaporize at atmospheric (flare) pressure, will end up mixed with the buffer fluid. There could then be two issues:

1) some pumped fluid, mixed with buffer fluid, would leak to atmosphere through the secondary seal. This many not be acceptable from a safety standpoint for toxic of highly flammable fluids.
2) The pumped fluid would accumulate in the seal pot.

When this can happen and is not acceptable, pressurized dual seals must be selected. The liquid medium in the secondary seal is called "barrier fluid," its purpose is to provide a barrier to prevent process liquid from leaking out to secondary seals.

Conversely, pressurized seals may be required for *highly volatile* pumped fluids; these are fluids prone to vaporize in the seal. Hence, they would not be suitable as sealing fluid. (API requires that the vapor pressure of the pumped fluid is above the seal chamber pressure with some margin. If this is not achieved, pressurized seals are required.)

In dual pressurized seals, the barrier fluid is kept under a pressure higher than that of the sealed chamber, allowing it to be used to flush the primary seal and blocking any leak of the pumped liquid.

Because some barrier fluid will leak into the process, it must be compatible with the process. The barrier fluid used/available at the site must be identified. Process design personnel must then verify that such contamination is not an issue. The seal system includes a pressurized pot, a pumping ring and a water or air-fin cooler. Pressure can be maintained by external means (nitrogen) or through a stand-alone pressurization system (bladder or piston inside the pot). A pressure transmitter is used to signal the loss of pressure.

API 682 provides informative (non-normative) guidelines for the selection of single/dual unpressurized or dual pressurized seals based on acceptability of leakage to atmosphere. Guidelines take into account personnel exposure, fire/explosion risk, impact on the environment, leakage detection and monitoring requirement. Commonly applied criteria are the following:

- Single seals shall be used wherever the choice satisfies safety, reliability, and ease of maintenance requirements in modern facilities.
- Dual unpressurized mechanical seals are specified for liquids that would release flammable or toxic vapors to atmosphere, i.e., highly volatile hydrocarbons such as LPG, liquids at a temperature above Auto Ignition Temperature (AIT) and below Initial Boiling Point (IBP), liquids containing H_2S.
- Dual pressurized seals are required for very dangerous liquids, e.g., acid, liquids at a temperature above Auto-Ignition Temperature and Initial Boiling Point (IBP), also for liquids containing more than 5% H_2S.

Make sure that the seal selection is based on sound reasoning because it influences cost and maintenance. Look for experience and good service in a seal vendor. Most seal vendors encourage the reader or user to download (or view) the 25 or 30 most frequently encountered API flush plans from the manufacturer's website.

The pump seal system, which comprises the mechanical seal cartridges themselves, equipment (pot, seal fluid cooler), piping, instruments (gauges, transmitters) is provided by the pump vendor and installed on the pump baseplate. The seal system should be specified to comply with API 682 for API 610 pumps; it could be per manufacturer's standard for other pumps, e.g., ASME pumps. However, note again that packing should no longer be used in pumps at facilities that wish to avoid leakage and/or do not wish to engage in the above-average maintenance requirements imposed by sealing solutions other than mechanical seals.

A.3.10.9 Barrier fluid

When pressurized dual seals are specified, the purchaser is well advised to specify the type of barrier fluid used at site, typically water or a mixture of water and industrial grade glycol. Automotive glycol is not suitable because these products contain additives that can attack certain seal components.

Request that the seal suppliers use industrial glycol unless it is not adequate. This will prevent the seal suppliers specifying proprietary barrier fluids which will increase the cost and make seal top-up operation, which is quite frequent (once a month) more complicated. The vendor must check that the proposed barrier fluid will not vaporize in the seal given the seal chamber pressure. The vendor must also ensure that the material of the gaskets (either Viton or FKM or FFKM) is compatible with the barrier fluid. Resist the temptation to use lubricating oils as a barrier fluid; oils will frequently not remove enough heat from seal parts.

The barrier fluid must be:
- Compatible with the pumped process fluids,
- Safe if leaking to atmosphere,
- First and foremost, acting as a heat removal agent. Its secondary function is providing a separation layer between the two seal faces. Acting as a quasi-lubricant for the seal should take third place.

A.3.10.10 Bearing lubrication

Understanding optimized pump lubrication is of great importance. Pump manufacturers fear that purchasers pay primary attention to bottom-line initial cost of pumps.

There is thus a disincentive to provide low-maintenance, highly reliable lube application methods. Reliability-focused owner-purchasers will specify superior lubrication methods[1].

Consider the next text segment an overview of readily available "old style" lubrication methods for pumps.

A.3.10.11 Oil bath

The simplest type of bearing lubrication is the oil bath. The oil, which is picked up by the rotating components of the bearing, is distributed within the bearing and then flows back to an "oil sump" in the housing. Typically, the oil level should almost reach the center of the lowest rolling element when the bearing is stationary.

The oil level is controlled by means of a "constant level lubricator." Some makes and models have been available since the year 1917. Most of these lubricators require servicing on a routine basis. There is a transparent external oil reservoir (an "oiler bulb") that maintains the level in the bearing housing. Vulnerabilities such as finite life of the caulking between oiler bulb and its metal support are not always brought to the user's attention.

Oil bath lubrication is particularly suitable for low speeds. It is adequate unless too much frictional heat is generated by the plowing action of the rolling elements at higher speeds. In practice, the oil bath lube method is suitable for pumps whenever DN (the inches of shaft diameter [D] multiplied by shaft revolutions per minute [N]) is below 6,000. See earlier chapters of this book for more detail.

A.3.10.12 Oil rings

Shafts with a DN > 6,000, for instance a pump with a two-inch shaft operating at 3,600 rpm, should use oil rings (sometimes misleadingly called "slinger rings") to lift the oil from a sump into the bearings. The oil level is then maintained below the bearing, which results in less frictional heat than would be generated if an oil bath would envelop the lower parts in a rolling element bearing. But again, if oil rings contact the surrounding bearing housing internals they tend to slow down or even abrade.

With modern synthetic lubricants and properly selected rolling element bearings, cooling is no longer needed in process pumps. Using high-performance synthetic lubricants and advanced bearing housing protector seals (Chapter 4) can extend oil and bearing life as much as ten-fold over older methods.

A.3.10.13 Forced oil lubrication

Only very large pumps need forced lubrication. These pumps incorporate plain (sleeve) bearings or tilting pad bearings which require oil to be at an optimum viscosity. Therefore, oil temperature control is needed. The forced lube oil system thus includes an oil supply pump, filter, and cooler.

When required, such an oil system could be specified to comply with API 610, which typically adds 25% to its cost by requiring that the reservoir and all oil piping be fabricated from austenitic stainless steel. The next increment is to specify API 614 which raises the system cost by a factor of 2.0 to 2.5. The extra cost is attributable to the larger reservoir (eight-minute retention time vs. three minutes), reservoir and oil piping fabricated from austenitic stainless steel, duplex full-capacity heat exchangers, more elaborate instrumentation, and extensive shop testing.

A lube-oil heater should be provided by the pump vendor if the site minimum ambient temperature is below the vendor-recommended minimum lube-oil temperature for safe start-up.

A.3.10.14 Oil mist

Oil mist lubrication is a standard practice in some industries; this practice is based on long-time and highly satisfactory experience. The superiority of oil mist is amplified by oil mist providing standby protection to non-running equipment. Substantial savings are attributed to extremely low maintenance requirements and lubricant re-use in closed oil mist systems. At all times, both pump and its electric motor driver are provided with oil mist. Payback for plant-wide systems is typically 1. 5 years.

Clean lubricants are carried to pump bearings in the form of oil mist. The mist is generated by centralized equipment, called an oil mist generator, wherein dry instrument-grade air is mixed with (typically) ISO VG 68 synthetic lube oil. The mixing ratio is 200,000 volumes of oil per volume of air; the resulting mist is distributed to all rolling element pump and motor bearings. Oil mist lubrication, if implemented, is applied to all pumps and drivers in green field projects. The CAPEX of an oil mist system is quite significant but plays a bigger role in savings to the OPEX side. An oil mist storage and preservation yard should be part of the CAPEX budget.

All pump and motor bearing housings are suitable for oil mist application. However, the relevant port locations must comply with guidelines found in API 610. Inlet/outlet port sizes, bearing type, inlet arrangement, etc. are described in books or can be obtained from oil mist providers. Some of these have been in business since the mid-1960s and can advise interested parties on oil mist system details and cost justification facts. Ref. 1 is the most up-to-date text on the subject.

A.3.10.15 Pump cooling

Despite friction encountered between moving parts and heat being generated as a result, pumps equipped with rolling element bearings and using synthetic oil do not need a cooling system. Heat dissipates by radiation from bearing housings to atmosphere. Beware of pump manufacturers whose bearing housings are designed and fabricated too small to function without cooling. These will be acceptable only with oil mist as the lubricant application strategy. In any event, if a vendor claims his pumps need cooling, API 610 requires the vendor to give reasons.

Tradition is often the reason for still finding new pumps incorporating a shaft driven fan. When the resulting flow of air is not enough, a cooling system is sometimes provided ("because that's what we always did"). It then cools the lubricating oil – regardless of need or no need. There are even instances where this is not enough and where traditionalists decide to circulate the cooling fluid through compartments "jackets" in the pump body.

Best-in-Class companies will not accept pumps with cooling jackets. Such pumps represent outdated designs which no longer harmonize with the environmental concerns of modern users. Moreover, cooling water jackets constrain free thermal expansion of a bearing. High preloading will result and rolling element bearings so cooled will fail prematurely.

For cooling of hydrodynamic or sleeve bearings, a needed cooling system includes a tank, pumps, and a cooler. It can be integrated into the equipment baseplate or installed as a separate lube oil console (skid). The cooler for hydrodynamic and sleeve bearings may use air or cooling water. Site conditions, availability, and conditions of cooling water must be disclosed to the pump supplier; this will enable the supplier/designer to properly engineer the desired lube oil system.

A.3.10.16 Space heater

A space heater or "anti-condensation" heater should be specified for electric motors in damp locations; this will prevent condensation of internal air during periods of shutdown. The heater maintains the temperature of the internal air above dew point; it should be activated when the motor is de-energized. It is usually specified for pumps above 75 kW. Oil mist systems can be furnished in lieu of space heaters.

A.3.10.17 PSVs

Reciprocating pumps must be fitted with a PSV (pressure safety valve) at their discharge in all cases. Centrifugal pumps do not usually require a PSV at their discharge. However, a PSV will be needed if the pump shut-off pressure exceeds the discharge line/equipment design pressure, which must be determined by the purchaser based on the shut-off head reported by the vendor for the pump(s) offered. To calculate the shut-off pressure the maximum suction pressure, i.e., suction vessel design pressure + head of liquid should be considered.

The discharge PSV on small reciprocating pumps, such as dosing pumps, is small enough to be mounted on the pump itself. It must be included in the scope of supply of the pump provider.

For large pumps, such as ones covered by API 674, the PSV is located off the skid. Its larger size makes it comparable to many other PSVs purchased by the purchaser and is best supplied by the owner/purchaser.

PSVs are sized for the maximum flow through the pump and the discharge design pressure. The vendor must size the driver for these PSV relief conditions.

A.3.10.18 Defining scope of supply

Reciprocating pumps, irrespective of size, must be supplied with pulsation dampeners at inlet and outlet to prevent pressure pulsations that can lead to pipe failure. These are sized by the vendors; accordingly, they must be included in the vendor's scope of supply. Use of inlet dampeners can be limited based on NPSHr and Acceleration Head Loss calculation: If the combination of NPSHr + Acceleration Head Loss is less than the NPSHa (without Acceleration Head Loss), then an inlet dampener can be avoided. In order for the vendor to properly calculate the acceleration head loss, the process data sheet should always indicate the available straight suction piping length. In addition, the NPSHa and suction pressure calculated by the Purchaser's Process Engineer and indicated on the process data sheet should be inclusive of all the losses from bends in suction piping and losses from any instruments or piping item such as valves or strainers.

A.3.10.19 Inlet strainers

A reciprocating pump's inlet and outlet valves are sensitive to particles; hence, they require inlet strainers. It will be better to leave these inlet strainers (Fig. A.3.14) in the vendor's scope of supply. However, this is not the case for centrifugal pumps. Their inlet strainers can be supplied by the purchaser whose standardization aims are justified for suction strainers. Suction strainers for centrifugal pumps are intended to catch construction-related debris. Best-in-Class companies usually remove and discard these strainers at the next opportunity.

Fig. A.3.14: Typical suction strainer.

A.3.10.20 Anchor bolt or setting bolt

Anchor bolts for pumps that are mounted on concrete foundation are commonly supplied by the purchaser's civil engineering team, but details of anchor bolts (number, size) should be defined by the pump vendor. For pumps mounted on a steel structure, the setting bolts are to be supplied by the vendor. This requirement also applies to all auxiliary skids in the total pump-related supply.

A.3.10.21 Hazardous area classification/explosion protection

The type of hazardous area, e.g., Zone 2 IIB T3, must be specified. The selection of the kind of explosion protection is best left to the vendor.

- "Ex na," in the future "ec," is the most cost-effective explosion protection; however, it cannot be used for motors in Zone 2.
- "Ex e," in the future "eb," motors are always expensive because they are de-rated to prevent hazardous temperatures, sparks and arcs. This de-rating makes "Ex e" unpopular and not widely used. When used with a variable speed drive, "Ex e" motors require additional testing. They must be tested together with the VSD as a complete system. This combined testing is not required for the other protection types.

- "Ex p" ratings are expensive for small motors but become less expensive on larger motors. In larger motors, the cost of the purge appurtenances becomes less in proportion to the overall cost.
- "Ex d," in the future "db," is the most expensive, especially for big motors for which the explosion proof enclosure becomes a larger proportion of the total motor cost. The tipping point is 1 MW: below 1 MW – "Ex d," above 1 MW – "Ex p."

In summary, in Zone 2, select "Ex na." In Zone 1, "Ex d" is probably the best choice, particularly if a VSD is used. However, in large sizes "Ex p" is the least expensive selection.

The applicable code for Explosion protection (IEC, CE ATEX) must be specified/ pointed out to the vendor. The pump supplier will specify the same to its sub-suppliers of electrical motor, instrumentation, etc.

A.3.10.22 Ingress protection

The suitable Ingress Protection (IP) of the electric motor as per IEC 60529 is the following:

- Indoor – IP 55
- Outdoor – IP 65 (perfectly waterproof) or, in presence of dust particles at the Plant, IP 66

The Ingress Protection Number of the instruments supplied with pumps must also be specified. IP 65 is the standard requirement for outdoor locations.

A.3.10.23 Instrumentation

The pump, motor and auxiliaries must be supplied with instrumentation for safe functioning. Instruments required for equipment protection should be provided by pump vendors as should instruments required for seal system, which shall be as per API 682.

Use of IP 65 or 66 mostly depends on the presence of dust particles at the plant, and whether the dust particles are conductive or non-conductive. The decision of which IP to be used should be made in consultation with the responsible instrumentation and electrical engineer.

1) Instruments required for explosion protection

Protection against excessive surface temperature, meaning exceeding the maximum limit of the class and for motors receiving variable input frequencies, temperature sensors must be embedded in the windings. These are RTD's (Resistance Temperature Detectors), commonly called Pt-100 because they are based on temperature sensitive resistors of 100 Ohms at 0°C. They are connected to the motor control center (MCC) and will trip a relay.

For motors without variable speed drive, these sensors are a protection against overload beyond the overcurrent-based overload protection already present in the MCC. They are therefore redundant and not always connected.

"Ex d" seal-less pumps require PTC Thermistor or PT 100 sensors for winding temperature monitoring. The responsible mechanical engineer should consult with an instrument engineer to identify which one should be specified to the vendor.

2) Instruments required for condition monitoring

Vibration instruments and temperature instrumentation per API 670 are provided to monitor the condition of large pumps only, typically above 500 kW. The client may also have requirements for vibration monitoring systems on pumps in specific service, for example LPG. Such instrumentation consists of accelerometers for vibration sensing, and RTDs, for temperature monitoring. These instruments are located at the bearings of both the pump and the motor. These sensors are normally connected to the Machinery Monitoring System, a dedicated system separate from the Process Control System.

Seal-less pumps often have more complex instrumentation, generally including detection of leaks past the containment boundary. Required instrumentation is specified in API 685.

A.3.10.24 Nozzle loads

Ask for twice the API 610 values as a reasonable standard strength requirement. For non-API pumps, select "by vendor" or "in compliance with ANSI/HI 9.6.2"

A.3.10.25 Inspections and tests

In the absence of client-specific requirements, the inspection and tests required by the code should be specified along with the following recommendations:

For all pumps purchased by and for reliability-focused plants, specify the following:

- Motor routine test
- Witnessed final assembly inspection (to verify workmanship by vendor and completeness before shipping)

The type of test should be requested for special (large) motors only.

For centrifugal pumps:

- Non-witnessed hydrostatic test for all pumps
- Non-witnessed performance and NPSHr test for all pumps
- Witnessed performance test for one item per service
- Witnessed NPIP/NPSH test for one item per service for services with NPSHa - NPSHr < 2 m

The following might be required for critical items only:

- Witnessed complete unit test, with job motor, seals, baseplate, for one item per service
- Witnessed mechanical running test (4 hours) for one item per service

API 674 pumps

- Witnessed mechanical and performance tests per API 674 for all items
- Witnessed NPIP/NPSH test for one item per service for services with NPSHa - NPSHr < 2 m

API 675

- Witnessed mechanical run test for all items
- Witnessed NPIP/NPSH test for one item per service for services with NPSHa - NPSHr < 2 m

A.3.10.26 Spare parts

Pre-commissioning spare parts usually include pump gaskets and O-rings. Bolts/nuts and gaskets for suction/discharge flanges should be requested if the pump is provided with counter or companion flanges.

However, this may be a fine opportunity to order upgraded spare parts for those pumps which, for any reason, were shipped with "traditional" parts. If improved, or upgraded parts are available, the spares should be upgraded versions. The two years of operation spare parts of a pump usually include one set of mechanical seals, pump gaskets, O-rings, bearings, and wear rings.

A quote for the two years spare parts should be requested from the vendor in the inquiry document. All parts not made by the pump manufacturer should be identified with originator's designation. Example: A particular bearing should not only be described as Bearing, XYZ Pump Mfr. ID 456386-5." Instead, the bearing should be called "Bearing, Universal Size ID 7216-G."

Capital or insurance spare parts include a spare impeller, which is commonly purchased for critical services only.

A.3.10.27 Vendor document delivery schedule

A pump set is considered packaged equipment, with the pump itself being usually the only part designed and manufactured by the pump vendor. The other parts, including seals, electric motor, coupling, lube oil console, if any, are sub-supplied (subcontracted) by the pump vendor. Therefore, even though the pump documentation could be submitted by the vendor early, obtaining documents for the packaged pump, complete with auxiliaries, takes time, usually around 70 days after PO.

It is critically important to ensure that the interface documents that enable engineering development (installation, operating and maintenance manuals) are received in time to allow site work. Among the required documents are foundation plan with loads, anchor bolt material description and number of bolts needed. It should be a contractual requirement for component details and size, utility consumption list, I/O list, P&ID, terminal box location) to be submitted in time to avoid installation delays. Penalty provisions should be written in the contract, perhaps in the form of finite percentages of the cost of equipment being withheld until critical documents are received by the purchaser or his EPC. Progress payments can be made contingent upon adherence to predefined delivery times.

Documents must be submitted at specified times, although that would rarely be earlier than eight to ten weeks after placing the Purchase Order. The first submission of the installation documents might indicate static loads (weight) only, while the second submission must indicate firmed-up dynamic and static loads to enable the design of foundations.

It is important that the supplier provide the equipment preservation procedures at the time of delivery. Oil mist preservation must be implemented to the fullest extent possible and old (usually outdated) beliefs reassessed by both the owner and his RPC contractor [1].

The following information should be provided in an acceptable time frame to the site installation team:

- Type of lubricant recommended for bearing housings, gearboxes, and motors; also, barrier or buffer fluid.
- Balancing instructions. Pump skids are usually balanced again at site during installation. Balancing instructions are usually available in IOM (installation, operation, maintenance manuals). While these instructions are usually quite straightforward for centrifugal pumps, they can be more difficult to obtain for

progressive cavity pumps. Accordingly, attention must be paid to this point and suitable procedures be obtained from the vendor.

- Grout and mass filler. Grouting quantity is estimated based on the foundation footprint multiplied by the grouting thickness. The baseplate cavity filler is included in the estimation, but the material should agree with end user practices. Multi-layer grouting shall be done as per API 686 recommended practice in the absence of client specific requirements. Do not overlook the advantages of using epoxy pre-filled baseplates (Refs. 2 through 4)

References

[1] Bloch, Heinz; Ehlert, Don; Geitner, Fred; "Optimized Equipment Lubrication, Oil Mist Technology, and Storage Preservation," Ft. Myers, FL., Reliabilityweb.com, 2020.

[2] Bloch, Heinz P. Petrochemical Machinery Insights. Oxford: Butterworth-Heinemann, 2016.

[3] Bloch, Heinz P. Pump Wisdom: Problem Solving for Operators and Specialists. Hoboken: John Wiley & Sons, 2011.

[4] Bloch, Heinz P. and Budris, Allan R. Pump User's Handbook: Life Extension. Fourth Edition. Lilburn, GA, Fairmont Press, 2013.

[5] Bloch, Heinz P. "Power End Upgrades Will Avoid Pump Failures." Proceedings of the 27th Texas A&M University International Pump Users Symposium, 2011.

[6] Bloch, Heinz P. and Lee, Buddy. "Breaking the Cycle of Pump Repairs." Proceedings of the 30th Texas A&M International Pump Users Symposium, 2014.

[7] Wilcock, Donald F. and Booser, E. Richard. Bearing Design and Application. New York City: McGraw-Hill Company, 1957.

Index

https://doi.org/10.1515/9783110674156-018

www.ingramcontent.com/pod-product-compliance
Lightning Source LLC
Jackson TN
JSHW060306061125
93675JS00003BA/117
* 9 7 8 3 1 1 0 6 7 4 1 3 2 *